Handbook of Valves and Actuators

Handbook of Valves and Actuators

Editor

Archana Sikarwar

Handbook of Valves and Actuators
Edited by **Archana Sikarwar**

Printed in 2017

ISBN: 978-1-68117-397-9

Library of Congress Control Number: 2015941587

© 2016 by
SCITUS Academics LLC,
616, Corporate Way, Suite 2, 4766,
Valley Cottage, NY 10989

www.scitusacademics.com

This book contains information obtained from highly regarded resources. Copyright for individual articles remains with the authors as indicated. All chapters are distributed under the terms of the Creative Commons Attribution License, which permits unrestricted use, distribution, and reproduction in any medium, provided the original author and source are credited.

Notice

Reasonable efforts have been made to publish reliable data and views articulated in the chapters are those of the individual contributors, and not necessarily those of the editors or publishers. Editors or publishers are not responsible for the accuracy of the information in the published chapters or consequences of their use. The publisher believes no responsibility for any damage or grievance to the persons or property arising out of the use of any materials, instructions, methods or thoughts in the book. The editors and the publisher have attempted to trace the copyright holders of all material reproduced in this publication and apologize to copyright holders if permission has not been obtained. If any copyright holder has not been acknowledged, please write to us so we may rectify.

Contents

Preface vii

Chapter 1   Electrorheology of Nanofiber Suspensions 1
                     Jianbo Yin and Xiaopeng Zhao

Chapter 2   Growth of Vertically Aligned ZnO Nanorods Using Textured ZnO Films 45
                     Francisco Solís-Pomar, Eduardo Martínez, Manuel F Meléndrez, and Eduardo Pérez-Tijerina

Chapter 3   Ammonium Fluoride-activated Synthesis of Cubic δ-TaN Nanoparticles at Low Temperatures 67
                     Young-Jun Lee, Dae-Young Kim, Kap-Ho Lee, Moon-Hee Han, Kyoung-Soo Kang, Ki-Kwang Bae, and Jong-Hyeon Lee

Chapter 4   Wettability Switching Techniques on Superhydrophobic Surfaces 85
                     Nicolas Verplanck, Yannick Coffinier, Vincent Thomy, and Rabah Boukherroub

Chapter 5   ELIXYS - A Fully Automated, Three-reactor High-pressure Radiosynthesizer for Development and Routine Production of Diverse PET Tracers 135
                     Mark Lazari, Kevin M Quinn, Shane B Claggett Jeffrey Collins, Gaurav J Shah Henry E Herman, Brandon Maraglia, Michael E Phelps, Melissa D Moore, and R Michael van Dam

Chapter 6   Quality of Electroless Ni-P (Nickel-Phosphorus) Coatings Applied in Oil Production Equipment with Salinity 165
                     Fernando B. Mainier, Maria P. Cindra Fonseca, Sérgio S. M. Tavares, Juan M. Pardal

Chapter 7	**Influence of the Chemical Composition of Completion Fluids on the Propagation of Electromagnetic Waves within Oil Wells** 187
	Alexandre Ashade Lassance Cunha, Marco Aurélio Pacheco, and José Ricardo Bergmann
Chapter 8	**No-moving-part Valve for Automatic Flow Switching**..................... 203
	Václav Tesař
	Citations.. 249
	Index 253

Preface

Valves are the components in a fluid flow or pressure system that regulate either the flow or the pressure of the fluid. They are used extensively in the process industries, especially petrochemical. Valves and actuators are widely used across industry and this dedicated reference provides all the information plant designers, specifiers or those involved with maintenance required. Industries that use pumps, seals and pipes will also use valves and actuators in their systems. This key reference provides anyone who designs, uses, specifies or maintains valves and valve systems with all of the critical design, specification, performance and operational information. In the presentations in their chapters, authors have written based on their own experience as well as those of their associates. Most chapters overlap several others. If differences of opinion or technique become apparent, seeming contradictions, it is because each author has had success with his or her particular technique, not that others are wrong. It should be recognized that each author's experience may be directed to a particular industry, perhaps omitting ideas important to other industries. An effort has been made to cross-reference chapters so that the reader may be aware of other points of view.d.

Editor

Chapter 1

Electrorheology of Nanofiber Suspensions

Jianbo Yin and Xiaopeng Zhao

Smart Materials Laboratory, Department of Applied Physics, Northwestern Polytechnical University, Xi'an 710129, China

ABSTRACT

Electrorheological (ER) fluid, which can be transformed rapidly from a fluid-like state to a solid-like state under an external electric field, is considered to be one of the most important smart fluids. However, conventional ER fluids based on microparticles are subjected to challenges in practical applications due to the lack of versatile performances. Recent researches of using nanoparticles as the dispersal phase have led to new interest in the development of non-conventional ER fluids with improved performances. In this review, we especially focus on the recent researches on electrorheology of various nanofiber-based suspensions, including inorganic,

organic, and inorganic/organic composite nanofibers. Our goal is to highlight the advantages of using anisotropic nanostructured materials as dispersal phases to improve ER performances.

INTRODUCTION

Since the discovery of carbon nanotubes (CNTs) by Iijima [1], there has been great interest in the synthesis, characterization, and applications of one-dimensional (1D) nanostructures. Nanofiber is an important class of 1D nanostructures, which offers opportunities to study the relationship between electrical, magnetic, optical, and other physical properties with dimensionality and size confinement. Various nanofibers including metal, inorganic, organic, and inorganic/organic composite have synthesized by different strategies [2-4]. Not only single nanofibers can act as building blocks for the generation of various nanoscale devices such as nanosensors, nanoactuators, nanolasers, nanopiezotronics, nanogenerators, nanophotovoltaics, etc. [5-14], but the incorporation of nanofibers in matrices would also produce advanced composite materials with enhanced properties [4,15-17]. On the other hand, due to some unique characteristics of nanofibers, such as small size, large aspect ratio, thermal, electronic, and transport properties, nanofiber-based suspensions or fluids have also received wide investigations for various applications in thermal transfer, microfluidics, fillers in the liquid crystal matrix, rheological, and biological fields [18-21].

Using external electric or magnetic fields to control the viscosity of fluids or suspensions is very interesting for science and technology because of the potential usage in active control of various devices in mechanical, biomedical, and robotic fields [22-24]. These fluids, whose viscosity can reversibly respond to external electric or magnetic fields, are often referred as 'smart fluids' which include liquid crystal, ferrofluid, magnetorheological (MR) fluid, and electrorheological (ER) fluid. ER fluid consisting of polarizable particles dispersed in a non-conducting liquid is considered to be one of the most interesting and important smart fluids [25, 26]. It can be transformed reversibly and rapidly from a fluid-like state to a solid-like state due to the disorder-order transition of particulate phase under an applied external electric field, showing tunable changes in the rheological characteristics. The tunable and quick rheological response to external electric fields makes ER fluid possess potential uses to enhance the electric-mechanical conversion efficiency in mechanical devices such as clutches, valves, damping devices, polishing, ink jet printer, human muscle stimulator, mechanical sensor, and so on [27-29]. In addition, some studies have shown that the ER fluid can be also used to fabricate potentially smart devices in optical, microwave, and sound fields [30-37].

The conventional ER fluid consists of micrometer-size dielectric particles in insulating liquid [25]. Since the ER effect was firstly discovered by Winslow [38], many ER systems including water-containing system such as silica gel, poly (lithium methacrylate), cellulose, and water-free system such as aluminosilicate, carbonaceous, semiconducting polymers have been developed. Some advanced materials including nanocomposites and mesoporous materials have also been investigated for ER fluid applications. The systematic introduction about the progress of ER materials, mechanisms, properties, and applications can be found in several literature reviews at different stages [39-52]. However, the present ER fluids do not possess a versatile performance, and there are still some disadvantages including insufficient yield stress, large particle settling, and temperature instability need to be overcome.

Some recent researches of using nanoparticles as the dispersal phase of ER fluid have led to new interest in the development of non-conventional ER fluid [53-56]. The nanopartile-based ER fluid exhibits extremely high yield strength though its large off-field viscosity and shear stability still need to be improved [57-61]. It is also interesting that compared with the suspension of spherical particles the suspension of 1D nanomaterials has been found to show some enhanced ER or MR effects and even improved dispersion stability recently. The present article provides a general overview on the electrorheology of nanofiber suspensions, including inorganic, organic, and inorganic/organic composite nanofibers.

Inorganic Nanofiber Suspensions

Although the effect of particle shape on ER properties has been noted for a long time [62, 63], one of the earliest experiments using elongated ER particles was reported by Asano et al. [64, 65]. They noted that the suspension containing both spherical and elongated particles produced the largest shear stress under an applied electric field. The suspension consisted of particles made of microcrystalline cellulose particles (The particle sizes were in the range of 20 to 400 µm.) dispersed in silicone oil. From microscopic observation, they suggested that spherical particles had a tendency to adhere to the electrodes, while elongated particles contributed to strengthening the particle chain. Kanu and Shaw [66] studied ER effect of a suspensions containing poly (p-phenylene benzobisthiazole) microfibres with different aspect ratios and found that the storage modulus increased significantly with the increase of aspect ratio. They attributed the increased ER effect to the overlapping of elongated particles and the increased dipolar interactions between elongated particles. Otsubo [67] also studied the effect of particle shape on ER effect by comparing the steady shear viscosity and oscillatory viscoelastic properties of whisker-like aluminum borate suspensions with spherical aluminum borate suspensions. The whisker sample had a diameter of 1 µm and a length of 30 µm, while the diameter of two spherical samples was 2 and 30 µm, respectively. Both steady shear viscosity and oscillatory viscoelastic experiments showed that the whisker suspensions showed a much higher ER response compared to the spherical suspensions at the same volume fraction. It was also found that when the stress amplitude was increased beyond the yield stress, the complex shear modulus of spherical aluminum borate suspensions showed a drastic decline due to the structural rupture. However, the complex shear modulus of whisker suspensions during oscillatory shear showed a shoulder-like decline after the stress exceeded the yield point [68]. The microscopic observation indicated that the fibrous column of whisker-like aluminum borate was thickened after oscillatory shear, which could well explain the enhancement of ER performances. Contrary to the results mentioned above, Qi and Wen [69] observed that the micro-sphere-based suspensions showed better ER performances than micro-rod-based suspensions when the particles had the same diameters. Based on the optical observation of chain-like structure, one possible reason they considered for this was that the micro-rods easily tangled together between the two parallel electrodes, and thus it was difficult for the micro-rods to align well in the direction of the external electric field. The tendency they found for the micro-rod-based suspensions was that the ER effect decreased with the increase of the aspect ratio, while this phenomenon became much weaker in the case when dried particles were substituted for the ones with moisture.

On the other hand, a particle level simulation model was reported recently for investigating the effects of elongated particles on the microstructure and field-induced flow response in the ER fluid [70]. The particles were modeled as a collection of spherical subunits joined by Hookean type connectors, which enabled the modeling of the particle motion through the Newtonian carrier liquid. The simulation results showed that the systems containing elongated particles possessed enhanced stress response when compared with those containing spherical particles at the same volume fraction, and this was similar to that observed from the experiments by Otsubo [67]. Furthermore, it was also pointed out that the stress contribution arising from rotational effects depended on the average orientation vector of the particles at the commencement of the shearing [70]. If the majority of the particles were tilted towards the direction of shearing, a positive contribution to stress would arise as a result of particles rotating against the direction of shearing towards the applied field direction.

Using inorganic nanofibers as the dispersal phase of ER fluid was firstly reported by Feng et al. [71]. In this report, ZnO nanowires were synthesized by thermal evaporation of Zn under controlled conditions without metal catalysts. The mean diameter of the nanowires was about 20 nm. The suspension was prepared by adding 1 g ZnO nanowires into 7 ml silicone oil and then manually stirring for about 30 min. Unlike the usual ER behavior, a decrease in viscosity (negative ER effect) for the ZnO nanowire suspension was observed under DC electric fields. According to the optical microscopic observation, such an anomalous behavior was considered to be due to the occurrence of the electrophoresis migration of ZnO nanowires to two electrodes induced by the electron transfer among ZnO nanowires. A positive ER effect of nanofiber suspensions was reported by the current authors by employing titanate nanofibers as dispersed phase [72, 73]. Titanate nanofibers were synthesized by a hydrothermal reaction of titania nanoparticles in high-concentration alkali solution following the Kasuga's report [74]. Titanate nanofibers were uniform nanotube-like morphology with outer diameter of 10 nm and length about 100-200 nm after ultrasonic (see Figure 1). High-resolution transmission electron microscopy (TEM) image (Figure 1d) and selected area electron diffraction (ED) (inset in Figure 1d) showed that the nanotubes consisted of the roll multilayered structure with an inner diameter of 3 nm. The energy-dispersive X-ray spectroscopy analysis showed the titanate nanofibers contained Na, Ti, and O elements. ER properties of suspension of titanate nanofibers in silicone oil were investigated by a steady shear viscosity. Compared to the suspension of titania nanoparticles, the suspension of nanofibers showed higher yield stresses (see Figure 2). At the same time, the alkali-ions intercalated in the interlayer of nanofibers were found to be important to the ER effect of titanate nanofibers. Removal of alkali-ions by acid-treatment did not destroy the nanofiber morphology (see Figure 1e) but weakened ER effect. According to the dielectric spectra analysis (see Figure 3), the decrease of ER effect was considered to be due to the degradation of dielectric property. However, it was noted that the ER effect of nanofiber suspension after removal of alkali-ions was higher than that of pure titania nanoparticle suspension. In particular, after 400°C calcination, the acid-treated nanofibers almost possessed the similar crystal structure and slightly higher dielectric constant compared with pure titania nanoparticles, but the ER effect of the former was still higher than that of the latter. This indicated that the anisotropic nanofiber structure played a role in improving the ER performance. In addition, the ER effect of titanate nanofiber suspension increased with increasing temperatures, which was in accordance with the improving dielectric properties. Another advantage of titanate nanofiber suspension was its lower particle settling rate compared to the conventional granular titania suspension.

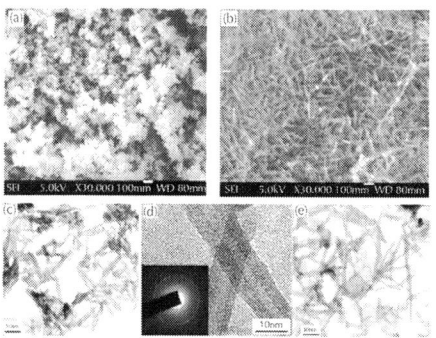

Figure 1: SEM and TEM images. SEM images of raw material of titania nanoparticles (a) and formed Na-titanate nanofibers after hydrothermal treatment and 250°C-annealing (b); low-magnification TEM (c) and high-resolution TEM and corresponding ED pattern (d) of Na-titanate nanofibers; (e) TEM image of formed H-titanate nanofibers by washing Na-titanate nanofibers with HCl solution [73].

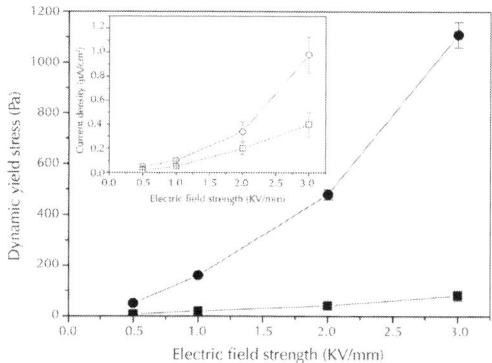

Figure 2: Yield stress as a function of electric field strength for Na-titanate nanofiber suspension (solid circle points) and titania nanoparticle sus-

pension (solid square points). The inset is the corresponding current density of Na-titanate nanofiber suspension (open circle points) and titania nanoparticle suspension (open square point) [72].

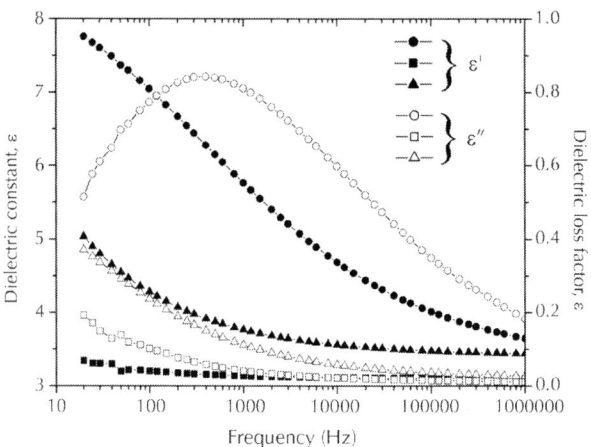

Figure 3: Dielectric spectra for the suspensions of titania nanoparticles (square points), 250°C-heated Na-titanate nanofibers (circle points), and 250°C-heated H-titanate nanofibers (triangle points) [73].

In order to investigate the changes of the microstructures of titanate nanofiber suspension under electric fields, the ER behavior of titanate suspension was further measured under oscillatory shear by He et al. [75, 76]. Investigation of ER properties by the dynamic oscillation method would be helpful to understand the nature of the interactions among particles forming the internal structures. The results showed that the dynamic moduli of titanate nanofiber suspension were much higher compared to original titania nanoparticle suspension under electric fields. Furthermore, the complex modulus of titanate nanofiber suspension was found to be sensitive to temperature, while that of titania nanoparticle suspension was insensitive at a higher temperature.

Lozano et al. [77] compared the ER effect of $Pb_3O_2Cl_2$ nanowire, carbon fiber (CNF), and single-walled CNT (SW-CNT) laden suspensions through oscillatory shear experiments in the presence of DC electric fields. It was observed that the CNF suspension developed a negative ER effect in which the storage modulus decreased with the increase of applied electric field. A decrease of 80% in storage modulus was observed at an electric field of 100 V/mm. In the case of the CNT suspension, a similar negative effect was observed. However, the $Pb_3O_2Cl_2$ nanowire suspension exhibited a positive ER effect and the maximum value was observed at 200 V/mm resulting in an increase of 120% in storage modulus. They considered that the observed negative ER effect in the CNF and CNT suspensions was related to the formation of a layered structure perpendicular to the direction of the electric field rather than a chain-like structure along the electric field direction, which was further due to the difference in electrical conductivity and polarization mechanisms.

Ramos-Tejada et al. compared the ER response of the suspension containing goethite (α-FeOOH) nanorods with axial ratio around 8 with the suspension containing polyhedral hematite (α-Fe_2O_3) particles with a mean diameter of 105 nm [78]. Both types of particles were said to possess similar chemical compositions and electrical properties and their average particle sizes were very close too. Thus, goethite and hematite samples differed mainly in particle shape. The experiments showed that the goethite suspension changed its rheological behavior from Newtonian without electric field to shear thinning at electric fields. In particular, the suspension of elongated goethite particles produced a more efficient ER response to the electric field than that made of polyhedral hematite particles since the former gave rise to higher yield stress for the same field strength, and exhibited a lower viscosity (see Figure 4) in absence of electric fields. As the chemical compositions and electrical properties, as well as the average particle sizes of elongated goethite and polyhedral hematite were very close, they attributed the ER enhancement to the larger dipole moments induced in elongated particles by the electric field. This consideration also justified why the goethite sample showed the same ER response as hematite one at low electric field of approximately 0.7 kV/mm, while their yield stresses differed significantly at high electric field of 1.5 and 2.0 kV/mm.

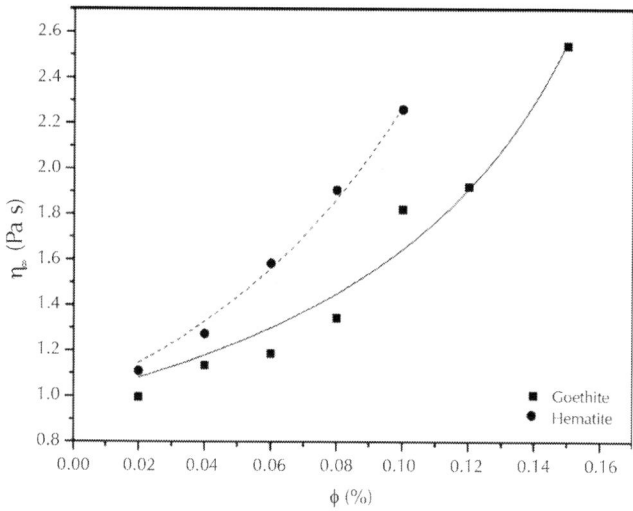

Figure 4: Viscosity at high shear rate as a function of the particle concentration for goethite and hematite suspensions. The lines correspond to the fit of the data to the Dougherty-Krieger equation [78].

A recent study by Cheng et al. [79] investigated the ER effect of a suspension of calcium and titanium precipitate (CTP) nanofibers. The nanofibers, which were prepared via a precipitation route in an ethanol/water mixed solution system containing tetrabutyl titanate, calcium chloride, oxalic acid dehydrate, had width of 23 nm and length of 40 to 130 nm (Figure 5). The nanofibers were claimed to be polycrystalline, but no clear crystal structure was ascertained according to the electron diffraction pattern. The X-ray diffraction pattern showed that the nanofibers were made of a complex mixture containing calcium oxalate dehydrate, $TiOC_2O_4(H_2O)_2$, and $TiO(OH)_2$. The rheological measurements showed that the complex nanofibers showed a large yield stress beyond 110 kPa at 66.6 wt% particle concentration in silicone oil, which was about twice higher as high as that of granular suspensions. From the absorption peaks at 3438 and 1649 cm^{-1} in Fourier transform infrared spectra, however, it could be judged that the nanofiber suspension belonged to a water-containing system. Therefore, the shortages of water effect on ER properties including thermal and electrical instabilities needed to be further overcome for the CTP nanofiber suspension.

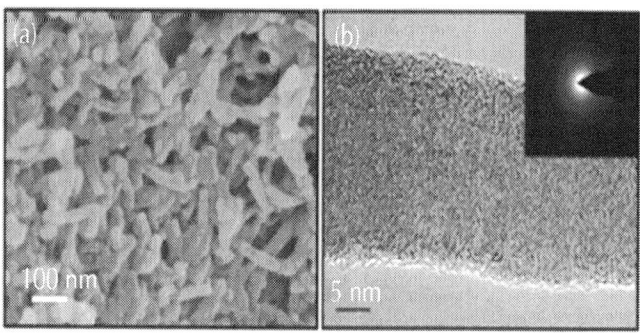

Figure 5: SEM image (a) and TEM image (b) with the SAED pattern in the inset of the calcium and titanium precipitate nanofibers [79].

Up to now, many kinds of inorganic nanofibers have been prepared by different techniques, but only amorphous or ionic crystal nanofibers can be used as high-performance ER fluids. Furthermore, the disadvantages including the large density and high abrasion of inorganic nanofibers need to be overcome.

Organic Nanofiber Suspensions

Due to low density and low abrasion to devices, organic ER systems have been widely investigated in the past decades. Polyelectrolytes and semi-conducting polymers are two kinds of important organic ER systems. In particular, the semi-conducting polymers including polyaniline (PANI), polypyrroles (PPy), poly(p-phenylene) (PPP), polythiophenes, poly(naphthalene quinine radicals) (PNQR), poly(acene quinine radicals) (PANQ), poly(phenylenediamine), oxidized polyacrylonitrile, and their derivatives have been frequently adopted as ER active materials because of the anhydrous character [45,47,49]. The interfacial polarization, induced by the local drift of electron or hole, is believed to be responsible for the ER effect of the semi-conducting polymer systems. By controllable adjustment of ϖ-conjugated bond structure, the conductivity and polarization can be changed.

Among these semi-conducting polymer ER systems, PANI has been considered as one of the most promising alternatives because of its simple preparation, low cost, good thermal stability, and controllable conduction and dielectric properties. Pure PANI and its modifications and composites have been developed for ER application in the past years [80-95]. Studies on these PANI materials greatly help the understanding about ER mechanisms and rheological properties. However, the application of ER fluids based on PANI is still limited to some extent by either low yield stress or particles, sedimentation.

Recently, one interesting way was developed to enhance the yield stress by employing nano-fibrous PANI [96]. The PANI nanofibers were easily synthesized on a large scale by an oxidative polymerization of aniline in an acid aqueous solution without mechanical stirring (see Figure 6). The outer diameter was of 200 nm and length of 1 to 5 µm. The BET surface area of PANI nanofibers was 43 m^2/g, which was higher than that (11 m^2/g) of granular PANI. After dedoping by immersion in 1 M aqueous ammonia, the PANI nanofibers with decreased conductivity were dispersed into silicone oil with grinding and ultrasonic to form suspensions. Compared to the conventional granular PANI suspension, the nanofiber suspension exhibited larger ER effect. Its shear stress and shear storage modulus were about 1.2 to 1.5 times as high as those of the former. At the same time, the shear stress of the PANI nanofiber suspension could maintain a stable level within the wide shear rate region of 0.1 to 1000 s^{-1} under various electric fields and the flow curves could be fitted by the Bingham fluid model (see Figure 7a). However, the shear stress of the granular PANI suspension showed a decrease as a function of shear rate to a minimum value, called the critical shear rate (see dot line in Figure 7b), after the appearance of yield stress and then increased again. The flow curves of Figure 7b could not be fitted by the simple Bingham fluid model but could be approximately fitted by the proposed Cho-Choi-Jhon model [97]. These indicated that anisotropic PANI nanofibers not only enhanced the yield stress but also influenced the flow behavior of suspension. In addition, it is interesting that the nanofiber suspension was found to possess better suspension stability compared to the conventional granular suspension when the particle weight fraction was same. No sedimentation occurred for the 15-wt% PANI nanofiber suspension after standing without disturbed for 500 h. This was considered to be related to the small size and large supporting effect of anisotropic nanofibers in suspensions [96].

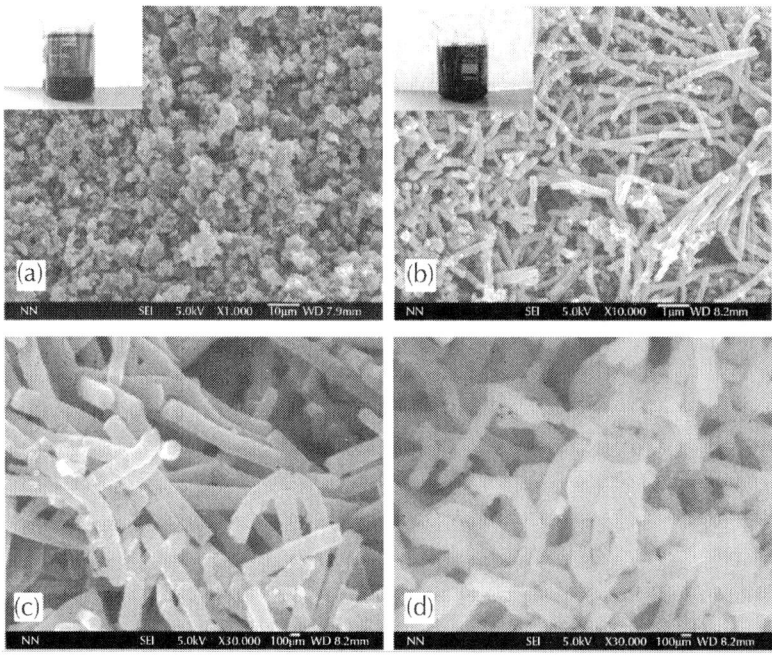

Figure 6: SEM images of samples: (a) granular PANI, (b) PANI nanofibers, (c) high resolution SEM images of PANI nanofibers, and (d) dedoped PANI nanofibers. The beakers shown in the insets contain the resultant granular PANI and PANI nanofiber suspensions, respectively [96].

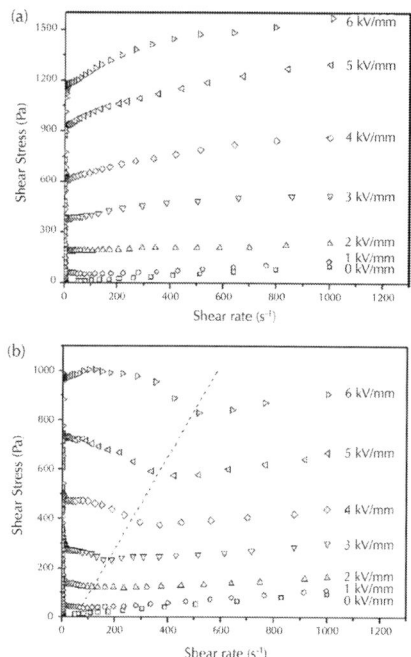

Figure 7: Shear stress as a function of shear rate for PANI suspensions under different DC electric fields: (a) nanofibers, (b) granular. (10 wt%, T = 23°C) [96].

By adjusting aniline/acid ratio or solution acidity, not only PANI nanofibers but also spherical micrometer-size and nano-size PANI particles were further prepared by a modified oxidative polymerization in low-cost citric acid solution and their electric, ER, sedimentation, and temperature properties were systematically compared recently [98]. It was found that the PANI nanofiber suspension exhibited the strongest ER effect under electric fields. Its yield stress was about 2.5 to 3.0 times as high as that of the PANI nanoparticle suspension and 1.3 to 1.5 times as high as that of the PANI microparticle suspension. The dependence of yield stress on electric field for the PANI nanofiber suspension was found to follow the power-law relation with a smaller exponent compared with the PANI nanoparticle suspension and microparticle suspension (see Figure 8). This was considered to be related to the anisotropic morphology of PANI nanofibers. The analogical result had also been obtained in the suspensions of spherical and whisker-like inorganic aluminum borate [67, 68]. Especially, it was interesting that the PANI nanofiber suspension was found to show lower off-field viscosity compared to the suspension of PANI nanoparticles, which proposed a possible way to overcome the problem of large off-field viscosity of the present nanoparticle-based ER fluids [57-61]. Furthermore, it was found that the PANI nanofiber suspension could maintain a good ER effect in a wide temperature range like the PANI microparticle suspension, while the temperature stability of the PANI nanoparticle suspension was degraded. It was known that the Brown motion disturbed ER structures in nanoparticle suspension systems more easily compared to microparticle suspension systems, but the larger dipole moments and more robust dendrite-like network induced by electric fields in PANI nanofiber suspension were believed to contribute to good temperature stability of ER effect [98].

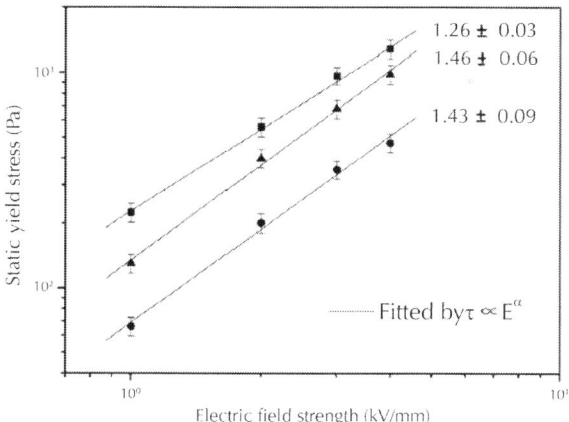

Figure 8: Static yield stress as a function of electric field strength (15 wt%, T = 23°C) for PANI suspensions: nanofibers (square points), microparticles (triangle points), and nanoparticles (circle points) [98].

Very recently, a kind of PPy nanofibers was synthesized for ER fluid application by a chemical oxidative polymerization and a thermo-oxidative treatment [99]. Under electric fields, the PPy nanofiber suspension possessed stronger ER effect than that of the conventional granular PPy suspension at the same volume fraction though the off-field viscosity of the former was lower than that of the latter. It also showed that the thermo-oxidative PPy nanofiber suspension could maintain good ER properties within a wide operating temperature range of 25 to 115°C.

Although organic nanofibers show more advantages in ER properties compared to the conventional granular ones, controlling the morphology of organic nanofibers in the preparation is more difficult compared to inorganic nanofibers. To extend the understanding about the effect of nanofiber morphology on ER properties, it is necessary to synthesize more kinds of organic nanofiber ER materials in the future works.

Carbonaceous Nanofiber Suspensions

Carbonaceous material is another very important kind of ER dispersal phase due to its anhydrous character, good ER efficiency, low density, and low electric power consumption. Carbonaceous ER material can be prepared from various organic sources [100-114]. For example, Kojima et al. [103,104] synthesized a kind of carbonaceous ER material composed of condensed polycyclic aromatic compounds with phenyl group and diphenyldiacetylene oligomers by annealing diphenyldiacetylene at an elevated pressure. Choi et al. studied the ER properties of pitch derived coke particles with different oxygen content or crystallographic properties [111]. Dong et al. [114] prepared the carbonaceous ER materials by thermal conversion of fluid catalytic cracking (FCC) slurry. Other carbonaceous materials have also been studied for use as the ER dispersant phase, including carbon black, graphitized carbon particles, carbon cones/disks, and mesoporous carbon[115-118].

CNTs have attracted a lot of scientific interest because of their anisotropic structure and outstanding electrical and mechanical properties for a wide range of applications [119]. In view of the unique characteristics of CNTs, in particular small size, large aspect ratio, thermal, and electronic properties, the ER properties of CNT suspensions have received wide investigations recently. Jin et al. [120] reported for the first time the ER properties of composites consisting of CNTs adsorbed polystyrene (PS) and poly-(methyl methacrylate) (PMMA) microspheres (see Figure9) when they were dispersed in silicone oil. The microscopic observation showed a clear chain structure formation in the suspension of CNTs adsorbed polymer microspheres when the external electric field was applied. After that, several kinds of composites containing CNTs were further developed by different techniques for ER fluid application [121-128].

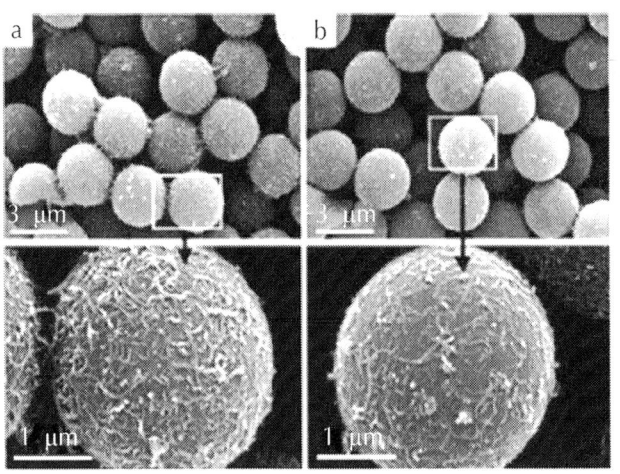

Figure 9: SEM images of the carbon nanotube-adsorbed PS microspheres using the surfactant: (a) CTAB and (b) NaDDBS [120].

Besides adsorbing onto the micospheres for ER fluid application, CNTs have also been added into ER and MR fluids as additives or fillers to decrease the serious particle sedimentation. For example, Fang et al. [129] have introduced SW-CNTs into carbonyl iron (CI) suspension as gap-filler to reduce the sedimentation of CI particles. Li et al. [130] have fabricated the ER fluid comprising nanoparticles/ multiwall CNTs (MW-CNTs) composite particles dispersed in silicone oil. This kind of ER fluid displayed dramatically enhanced anti-sedimentation characteristic compared to the ER fluid without MW-CNTs. In the best cases, stabilized suspensions after adding MW-CNTs have been maintained for several months without any appreciable sedimentation being observed. The addition of MW-CNTs was considered to introduce an effective short range repulsive interaction between the ER nanoparticles. However, such repulsive interaction only slightly decreased the yield stress under an electric field.

Although adding CNTs into conventional ER or MR fluids has improved the suspension stability, CNTs only act as fillers or additives in these studies. The alignment and polarizability of pure SW-CNT suspensions under electric fields have been investigated through optical polarimetry by Brown et al. [131]. In the study, a low-frequency alternating-current electric field was applied and the nematic order parameter was determined by measuring changes in the state of polarization of a laser beam transmitted through the suspension. They found that the dependence of the measured alignment of SW-CNTs on the electric field was consistent with a thermal-equilibrium distribution of freely rotating, polarizable rods. The polarizability determined by fitting to this model was consistent with the classical result for a conducting ellipsoid of the dimensions of the nanotube. Recently, Lin et al. [132] further measured the apparent viscosity of a dilute SW-CNT/terpineol suspension under an external electric field. Although the volume fraction of SW-CNTs was very small of 1.5×10^{-5}, it was experimentally found that the viscosity of suspension increased to more than double at moderate shear rates and electric field of 160 V/mm. In particular, they observed the magnitude of the ER response in the dilute SW-CNT suspension was much higher than that of the conventional suspension containing micro-size glassy carbon spheres at comparable volume fractions. For the suspension of glassy carbon spheres, a suspension of, a three-order-of-magnitude-higher volume fraction must be required to achieve similar increases in the apparent viscosity under the same conditions. The ER response of SW-CNT suspension could be interpreted in terms of an electrostatic-polarization model and the enhanced ER response was attributed to the improved polarization and drag force due to high aspect ratio of the CNTs. Furthermore, the ensemble-averaged particle-orientation angles and apparent shear viscosities of dilute suspensions of SW-CNT/terpineol were also experimentally studied by an optical polarization-modulation method under electric fields during flow recently [133]. Particle-orientation angles for various shear rates (D) and electric fields (E) were found to collapse when plotted against the parameter, f~ E^2/D as predicted by the theory developed by Mason and co-workers for the equilibrium orientation angle of ellipsoids under electric fields and shear flow. However, comparison between measured and predicted particle-orientation angles showed poor agreement at intermediate values of f. Electrostatic interactions among large-aspect-ratio particles were shown to be significant, and might account for the discrepancy between the measurements and classical theories for even dilute suspensions of nanotubes under both shear and electric fields. Under DC electric fields, however, the CNT suspension showed a negative ER behavior due to large electrical conductivity [77].

The CNT suspensions mentioned above are made of the commercial CNTs, their yield strength or ER efficiency is too low to be used in many ER devices and the electrical breakdown easily occurrs in these suspensions containing commercial CNTs because of the easy percolation of pseudo-1D conductivity [77,132].

Very recently, a kind of nanotube-like nitrogen-enriched carbonaceous nanofibers (N-CTs) were prepared by the heat treatment of conducting PANI nanofibers and then were used as new carbonaceous ER materials [134]. The heat treatment temperature was found to be important to obtain N-CTs with the optimal ER effect. The heat treatment at the temperature lower than 500°C easily transformed PANI nanofibers into thermally degraded PANI nanofibers whose conductivities were too low to induce a strong ER effect, while the heat treatment at temperature higher than 600°C transformed PANI nanofibers into the partially graphitized nitrogen-containing nanotubes whose conductivities were too high to finish ER measurements because of the electrical short circuit.

When PANI nanofibers were treated in vacuum at the temperature range of 500 to 600°C, the obtained N-CTs were suitable to be used as ER dispersal phase because they had the moderate conductivity. After heat treatment, the nanofiber morphology was found to be well preserved except that the diameters showed shrinkage and the aspect ratio of nanotubes slightly decreased with increasing heat treatment temperatures [134]. Figure 10 showed the morphology and Raman spectra of N-CTs obtained at 550°C. The N-CTs possessed the uniform nanotubular morphology with a diameter of 90 to 150 nm and a length of 1 to 2 μm. The Raman spectra of the N-CTs showed two broad bands centered at about 1588 cm^{-1} (G band) and 1345 cm^{-1} (D band), characteristic of amorphous carbon or disordered graphites. The N-CTs mainly contained C (77.5 wt %), N (12.6 wt %), and other elements (such as H and O).

These indicated that the heat treatment at 550°C had transformed the PANI nanofibers into the amorphous nitrogen-enriched carbonaceous nanotubes [135]. Under electric fields, the rheological results showed that the N-CT suspension possessed versatile ER performance including high ER efficiency, good dispersion stability, and temperature stability. Especially, compared to the corresponding suspension of heat treated granular PANI, the N-CT suspension showed better dispersion stability and higher ER effect (see Figure 11). The analogical result was also observed in the dilute ER fluid containing commercial CNTs [132]. When a power-law relation $\tau_y \propto E$ was used to fit the correlation of yield stresses and electric fields, it was also found that the exponent of the N-CT suspension was smaller than that of granular suspension. This was mainly related to the particle morphology because other factors such as particle concentration, particle's conductivity, liquid phase, and so on were the same for N-CTs and heat treated granular PANI. The similar result was also observed in the PANI nanofiber suspension [96, 98] and in the whisker-like inorganic aluminum borate suspension [67]. Furthermore, the ER effect of N-CT suspension could be adjusted by varying heat treatment temperatures and the N-CTs obtained at around 600°C exhibited the maximum ER effect (see Figure 12). This was explained by the polarization response, which originated from the regular change of conductivity of N-CTs as a function of heat treatment temperatures [134].

It showed that under electric fields the N-CT suspension showed good temperature stability in ER effect though its off-field viscosity decreased with elevated temperatures. Meanwhile, the flow curve of shear stress vs. shear rate also maintained a stable level and the critical shear rate shifted toward high values as the operating temperature increased. The dynamic viscoelastic measurement showed that the storage modulus slightly increased with increasing operating temperature, also confirming the good temperature stability of ER effect of N-CT suspension. The dielectric spectra of N-CT suspension and the dielectric parameters calculated by the Cole-Cole equation could explain the temperature dependence of ER effect of N-CT suspension [135].

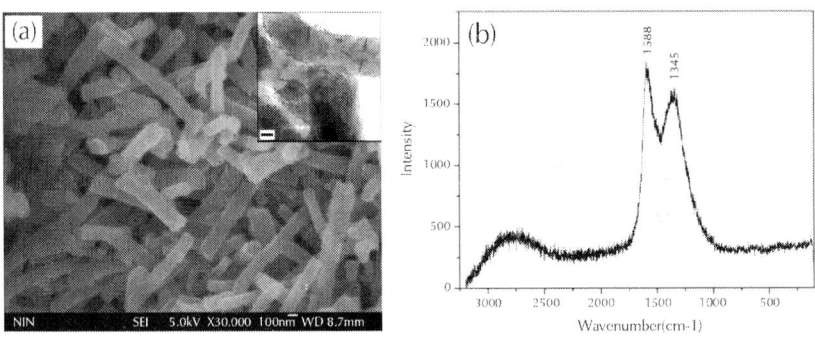

Figure 10: The morphology and Raman spectra of N-CTs. (a) SEM image and TEM image (inset, scale bar = 50 nm) of N-CTs, (b) Raman spectra of N-CTs [135].

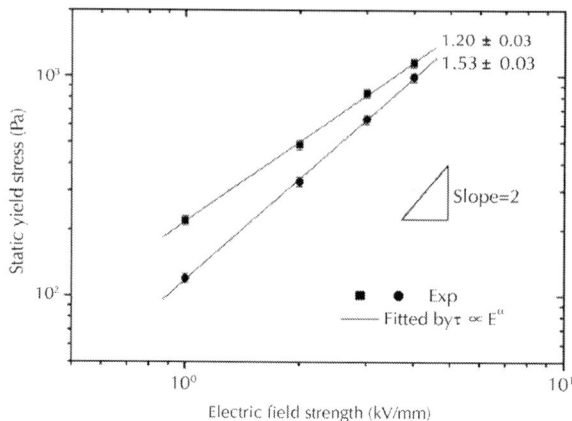

Figure 11: Yield stress as a function of electric field strength for N-CT suspension (square symbol) and heat treated granular PANI suspension by the same process (circle symbol) (T = 23°C, 15 vol.%) [134].

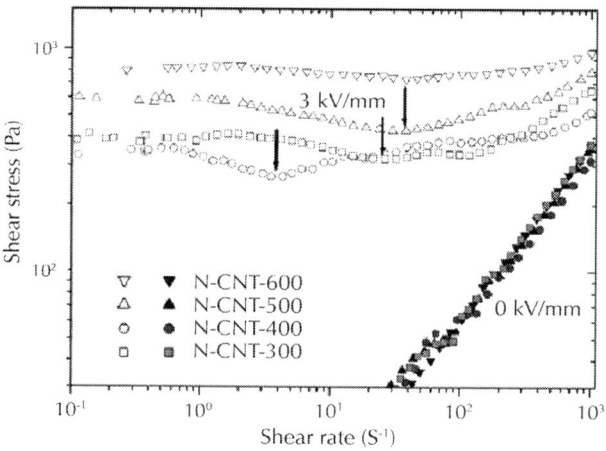

Figure 12: Flow curves of shear stress vs. shear rate for N-CT suspensions under zero (solid symbol) and 3 kV/mm (open symbol) electric fields (T = 23°C, 15 vol.%) [134].

The field response of vapor-grown carbon nanofibers (VGCFs) was also observed when dispersed in polydimethylsiloxane [136]. It was found that a DC electric or magnetic field was applied to induce the formation of an aligned structure. Upon application of a DC electric field, an aligned ramified network structure of VGCFs developed between the electrodes. In the formation of the network structure, ends of VGCFs became connected to ends of other VGCFs, which were followed by rotation and orientation of the VCGFs. On the other hand, upon application of a magnetic field, the VGCFs were only rotated, without the formation of a network. The viscosity of the polydimethylsiloxane matrix was found to influence the structural formation process. However, no rheological data were reported in the VGCFs/polydimethylsiloxane suspension.

Although 1D carbonaceous material is potential as novel nanofiber ER fluids, it should point out that the suspension durability or dispersion stability is still a challenge due to the facile aggregation of 1D carbon nanomaterial. One feasible way of improving dispersion stability is to prepare the polymer graft 1D carbonaceous material by the graft reaction of carboxyl groups on the carbon material [137].

Inorganic/Organic Composite Nanofiber Suspensions

Although the inorganic and organic ER materials show many advantages, the disadvantages of single component are also prominent and difficult to be harmonized. To obtain ER fluids with comprehensive performances, the fabrication of composite ER particles have been proposed because they can combine the advantages of different components. The most popular composite ER particles are core/shell structured particles [138-142]. On one hand, the particle sedimentation problem of ER fluids is expected to be overcome by using low density polymer or hollow sphere as core. On the other hand, it is considered to be feasible to increase ER effect by adjustably controlling the conductivity and dielectric constant of core and shell. The detailed theorized investigations have included various core/shell composite particles [143-147]. It has indicated that promising ER fluids for using over a wide frequency range were those which contained highly conducting particles coated with an insulating shell having high dielectric constant and high electric breakdown strength.

Having considered the advantages of core/shell composite ER particles, researchers have paid significant attention to inorganic/organic composite nanofibers for use as promising ER fluids recently [148-152]. By combining conducting polymer nanofibers and insulating inorganic dielectric, a kind of well-organized coaxial cable-like PANI@titania nanofibers was synthesized by a facile hydrolysis of tetrabutyl titanate in the presence of conducting PANI nanofibers for ER fluid application [148]. From Figure 13, it was noted that the PANI nanofiber core had a diameter of 150 to 200 nm and length of 0.5 to 3.0 μm. The thickness of sheath layer was about several tens of nanometers, depending on the amount of water used in the reaction. The sheath thickness increased with the increase of the amount of water. In the coating process of titania sheath, the chemical structure of PANI nanofibers was less changed because of no additional acid or alkali was added and thus the physical properties of PANI core was expected to maintain unchanged. Under electric fields, the suspension of PANI@titania nanofibers exhibited the lightly smaller yields stress compared to that of the suspension of pure dedoped PANI nanofibers, but its leaking current density was significantly lower than that of the latter. This was attributed to a sufficient electrical insulating effect of titania sheath to conducting PANI core (the dielectric constant of the PANI@titania nanofiber suspension was about 5.0 at 1 kHz) and thus the coaxial cable-like PANI@titania nanofibers could be used as a potential dispersal phase of ER fluid with low electric power consumption.

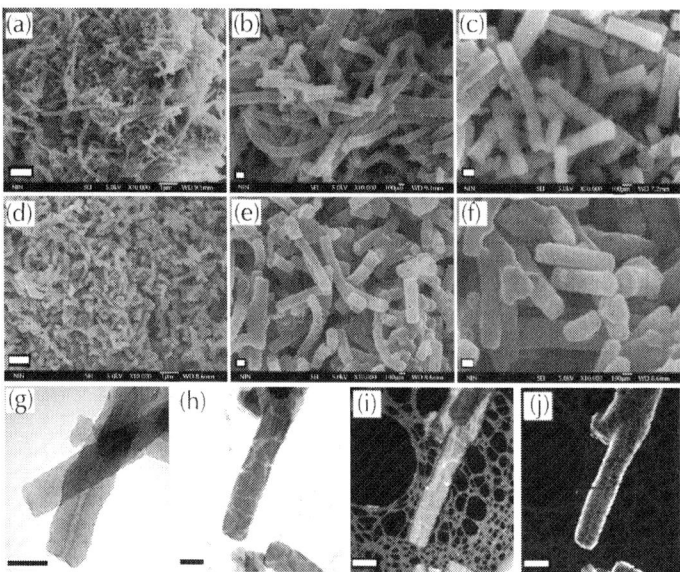

Figure 13: SEM images: as-made PANI nanofibers (a to c) and PANI@titania nanofibers (d to f). (The scale bar is 1 μm for (a) and (d), 100 nm for (b), (c), (e), and (f)). TEM images: as-made PANI nanofibers (g) and PANI@titania nanofibers (h). EELS analysis of PANI@titania nanofibers for C element (i) and Ti element (j). (The scale bar is 200 nm for (g) to (j)) [148].

A silica nanoparticle decorated PANI nanofibers were also successfully synthesized as a dispersed phase of an ER fluid recently [149]. In this study, the PANI fibers obtained through interfacial polymerization were about 300 to 400 nm in diameter and 2 to 5 μm in length. Then the fibers were redispersed in ethanol containing tetraethyl orthosilicate (TEOS), and silica nanoparticles were formed on the surface of the fibers through a modified Stöber method (see Figure 14). Due to the use of the hydrous ammonia in the synthesis, however, the PANI fiber core was the dedoped emeraldine base-form in the resulted silica nanoparticle decorated PANI fibers. The ER properties of the suspensions based on pure PANI fibers and silica-PANI fibers were compared using a rotational rheometer under electric fields, demonstrating lower shear stress and slight different flow curves for the silica decorated PANI fiber suspension.

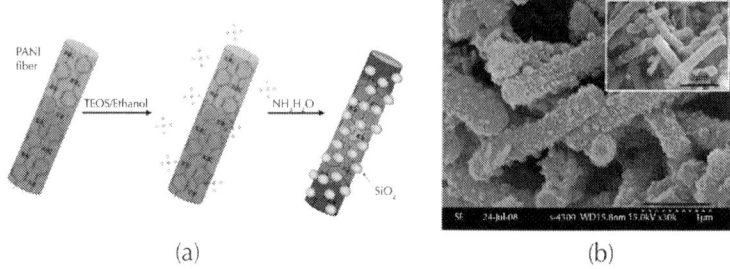

Figure 14: Silica-PANI fibers. (a) Schematic diagram of synthetic process of silica-PANI fibers and (b) SEM images of resulted silica-PANI fibers (inset: PANI fibers) [149].

On the other hand, the composite nanofibers composed of inorganic core coated by organic sheath were also developed for ER fluid application. Cheng et al. [150] have synthesized PANI/titanate composite nanofibers by in situ chemical oxidative polymerization directed by block copolymer. In their preparation, the inorganic titanate nanofibers were modified first by block copolymer and then PANI was coated by chemical oxidative polymerization of aniline monomer in the presence of modified titanate nanofibers. Although the authors did not give a comparison of ER effect of the composite nanofibers with the single core or shell component, it was found that the ER activity of PANI/titanate nanofiber suspension varied with the ratio of aniline to titanate. In particular, the PANI/titanate nanofiber suspension was found to show a higher ER effect than that of the sphere-like PANI/titania nanoparticle suspension, indicating a positive contribution to ER enhancement from the anisotropic morphology. The ER enhancement was interpreted by the dielectric spectra analysis; a larger dielectric loss enhancement and a faster rate of interfacial polarization were responsible for the higher ER activity of the PANI/titanate nanofiber-based suspension. It should be pointed out that, different from the cable-like PANI@titania nanofibers mentioned above, the PANI/titanate composite nanofibers must be dedoped to decrease the conductivity of PANI sheath before they were used as ER dispersal phase. Otherwise, the suspensions will subject to an electrical short circuit under high electric fields. Compared to those composed of organic core coated with inorganic sheath, however, this kind of inorganic/organic composite nanofibers possessed an advantage of low abrasive action to devices.

Core/shell composite nanofibers can act as the model materials to match the advantages of different components for the optimal ER performances, but the wreck of coating layer under high shearing force is still a problem to limit their practical applications. The formation of inorganic/organic hybrid composite nanofibers provides an alternative way. Due to the stronger interaction between inorganic and inorganic components, the hybrid composite nanofibers are expected to possess more stable mechanical properties when the suspensions subjected to strong shearing flow. For example, a kind of conducting PPy nanofibers with TiO_2 nanoparticles was synthesized in the presence of -naphthalenesulfonic acid by chemical oxidative polymerization recently [151]. The results indicated that the structural and electrical properties of the composite nanofibers were influenced by the content of TiO_2 nanoparticles. The DC conductivity of the composite nanofibers increased by one order of magnitude when the concentration of TiO_2 was 0.1 M compared with pure PPy nanofibers. The AC conductivity of the composite nanofibers showed the similar trend with the TiO_2 content and obeyed the power law index in the 10 to 10^7-Hz range. The ER properties of the composite nanofibers in silicone oil were also evaluated under steady and oscillatory shear. Chuangchote et al. used an electrospinning method to fabricate mats of nanofibers from neat and carbon black (CB) nanoparticle-loaded poly(vinyl alcohol) (PVA) solutions in reverse osmotic water. The ER properties of the as-spun neat and CB-loaded PVA fiber mats with the average diameter of the individual fibers being about 160 nm and the thickness of the mats being about 20 to 30 μm were characterized [152]. While their Young›s modulus was found to increase, both the tensile and the elongation at break of the as-spun fiber mats were found to decrease, with the addition and increasing amount of CB. With or without the electrostatic field, both the storage and the loss moduli for all of the as-spun neat and CB-loaded PVA fiber mats were found to increase with increasing frequency. Under the electrostatic field, the dynamic mechanical responses were found to increase with initial increase in the applied electrostatic field strength (EFS) and level off at a certain applied EFS value. At the applied EFS value of 100 V/mm, the dynamic mechanical responses were found to increase with the initial increase in the CB content and level off when the CB content was greater than about 6%. However, no viscosity properties were studied for the CB-loaded PVA naonfibers when dispersed in insulating liquid.

CONCLUSIONS

The preparation of non-conventional ER fluids based on nanoparticles is an area of growing interest from both the fundamental and application points of view. In this review, we have summarized recent researches in the synthesis and ER properties of nanofiber suspensions including inorganic, organic, and inorganic/organic composite nanofibers. Although Qi and Wen [69] have observed that the microsphere based suspensions showed high yield stress than that of micro-rod based suspensions when the particles had the same diameters, most of researches have indicated that the small size and anisotropic structure with large aspect ratio played a great role in improving the suspension stability and ER properties of nanofiber suspensions compared to the conventional sphere suspensions. Some nanofiber suspensions have also been found to show lower off-field viscosity compared to nanosphere suspensions, which provides a possible way to solve the problem of large off-field viscosity of present nanoparticle based ER fluids. Especially, it should be noted that the theoretical and experimental investigations performed recently on MR fluids also showed that suspensions containing magnetic fibers or nanofibers gave rise to an enhanced magnetorheology when compared with conventional MR fluids made up of spherical particles [153-163]. Therefore, it is reasonable to point out that employing anisotropic nanostructured particles to improve ER performances is a very interesting topic. However, the disadvantages including complicated preparation process, nanofiber aggregation, etc. and the further understanding about physical and chemical mechanisms behind the electrorheology of nanofiber suspensions need to be noted in the future works. In addition, the exploration of the nanofiber ER suspensions for new applications in advanced sensors and actuators in MEMS and biotechnology fields should be noted.

ACKNOWLEDGMENTS

The financial support from the National Natural Science Foundation of China (no. 50602036, 50936002) and the NPU Foundation for Fundamental Research (no. JC201051) is acknowledged.

REFERENCES

1. Iijima S: Helical microtubules of graphitic carbon. Nature 1991, 354:56.
2. Xia YN, Yang PD, Sun YG, Wu YY, Mayers B, Gates B, Yin YD, Kim F, Yan YQ: One-dimensional nanostructures: Synthesis, characterization, and applications. Adv Mater 2003, 15:353.
3. Rao CNR, Deepak FL, Gundiah G, Govindaraj A: Inorganic nanowires. Prog Solid State Chem 2003, 31:5.
4. Huang ZM, Zhang YZ, Kotakic M, Ramakrishna S: A review on polymer nanofibers by electrospinning and their applications in nanocomposites. Compos Sci Technol 2003, 63:2223.
5. Kind H, Yan HQ, Messer B, Law M, Yang PD: Nanowire ultraviolet photodetectors and optical switches. Adv Mater 2002, 14:158.
6. Johnson JC, Choi HJ, Knutsen KP, Schaller RD, Yang PD, Saykally RJ: Single gallium nitride nanowire lasers. Nat Mater 2002, 1:106.
7. Zheng GF, Patolsky F, Cui Y, Wang WU, Lieber CM: Multiplexed electrical detection of cancer markers with nanowire sensor

arrays. Nat Biotechnol 2005, 23:1294.
8. Wang ZL, Song JH: Piezoelectric nanogenerators based on zinc oxide nanowire arrays. Science 2006, 312:242.
9. Tian BZ, Zheng XL, Kempa TJ, Fang Y, Yu N, Yu G, Huang J, Lieber CM: Coaxial silicon nanowires as solar cells and nanoelectronic power sources. Nature 2007, 449:885.
10. Zhou J, Gu YD, Fei P, Mai WJ, Gao YF, Yang RS, Bao G, Wang ZL: Flexible piezotronic strain sensor. Nano Lett 2008, 8:3035.
11. Garnett EC, Yang PD: Silicon nanowire radial p-n junction solar cells. J Am Chem Soc 2008, 130:9224.
12. Yang RS, Qin Y, Dai LM, and Wang ZL: Power generation with laterally packaged piezoelectric fine wires. Nat Nanotechnol 2009, 4:34.
13. Tian BZ, Kempa TJ, Lieber CM: Single nanowire photovoltaics. Chem Soc Rev 2009, 38:16.
14. Wang ZL: Piezotronic and piezophototronic effects. J Phys Chem Lett 2010, 1:1388.
15. Sazonova V, Yaish Y, Ustunel H, Roundy D, Arias TA: A tunable carbon nanotube electromechanical oscillator. Nature 2004, 431:284.
16. Cao FF, Guo YG, Zheng SF, Wu XL, Jiang LY, Bi RR, Wan LJ, Maier J: Symbiotic coaxial nanocables: Facile synthesis and an efficient and elegant morphological solution to the lithium storage problem. Chem Mater 2010, 22:1908.
17. Eder D: Carbon nanotube-inorganic hybrids. Chem Rev 2010, 110:1348.
18. Xie HQ, Lee H, Youn W, Choi M: Nanofluids containing multiwalled carbon nanotubes and their enhanced thermal conductivities. J Appl Phys 2003, 94:4967.
19. Dierking I, Scalia G, Morales P: Liquid crystal-carbon nanotube dispersions. J Appl Phys 2005, 97:044309.
20. Garcia AA, Egatz-Gomez A, Lindsay SA, Melle S, Marquez M, Domínguez-Garcia P, Rubio MA, Picraux ST, Yang D, Aella P,

Hayes MA, Gust D, Loyprasert S, Vazquez-Alvarez T, Wang J: Magnetic movement of biological fluid droplets. J Magn Magn Mater 2007, 311:238.
21. Bauhofer W, Schulz SC, Eken AE, Skipa T, Lellinger D, Alig I, Tozzi EJ, Klingenberg DJ:Shear-controlled electrical conductivity of carbon nanotubes networks suspended in low and high molecular weight liquids. Polymer 2010, 51:5024.
22. Tsuchiya K, Orihara Y, Kondo Y, Yoshino N, Ohkubo T, Sakai H, Abe M: Control of viscoelasticity using redox reaction. J Am Chem Soc 2004, 126:12282.
23. Han YM, Kang PS, Sung KG, and Choi SB: Force feedback control of a medical haptic master using an electrorheological fluid. J Intell Mater Syst Struct 2007, 18:1149.
24. Sim ND, Stanway R, Johnson AR, Peel DJ, Bullough WA: Smart fluid damping: shaping the force/velocity response through feedback control. Smart Materials and Structures 2000: Smart Structures and Integrated Systems 2000, 3985:470. 0-8194-3603-8
25. Halsey TC: Electrorheological fluids. Science 1992, 258:761.
26. Wen W, Huang X, Sheng P: Electrorheological fluids: structures and mechanisms. Soft Matter 2008, 4:200.
27. Block H, Kelly JP: Electrorheology. J Phys D Appl Phys 1988, 21:1661.
28. Coulter JP, Weiss KD, and Carlson JD: Engineering applications of electrorheological materials. J Intell Mater Syst Struct 1993, 4:248.
29. Hao T: Electrorheological fluids. Adv Mater 2001, 13:1847.
30. Zhao Q, Zhao XP: Tunable optical activity in electrorheological fluids. Phys Lett A 2005, 334:376.
31. Zhap Q, Zhao XP, Qu C, Xiang LQ: Diffraction pattern and optical activity of complex fluids under external electric field. Appl Phys Lett 2004, 84:1985.
32. Fan JJ, Zhao XP, Gao XM: Electric field regulating behavior of microwave propagation in ER fluids. J Phys D Appl Phys

2002, 35:88.
33. Zhao XP, Huang M, Tang H: Trough and hump phenomenon in a kind of random system. Phys Lett A 2005, 339:159.
34. Tang H, Luo CR, and Zhao XP: Tunable characteristics of a flexible thin electrorheological layer for low frequency acoustic waves. J Phys D Appl Phys 2004, 37:2331.
35. Tang H, Luo CR, Zhao XP: Sonic resonance in a sandwiched electrorheological panel. J Appl Phys 2005, 98:016103.
36. Tang H, Luo CR, and Zhao XP: Sonic responses of an electrorheological layer with one side of grating electrodes. J Phys D Appl Phys 2006, 39:552.
37. Tang H, Zhao XP, Wang BX, Luo CR: Respose characteristics of a viscoelastic gel under the co-action of sound waves and an electric field. Smart Mater Struct 2006, 15:86.
38. Winslow WM: Induced fibration of suspensions. J Appl Phys 1949, 20:1137.
39. Conrad H, Sprecher AF: Characteristics and mechanisms of electrorheological fluids. J Stat Phys 1991, 64:1073.
40. Weiss KD, Coulter JP, Carlson JD: Electrorheological fluid under elongation, compression and shearing. J Intell Mater Syst Struct 1993, 4:13.
41. Gast AP, Zukowski CF: Electrorheological fluids as colloidal suspensions. Adv Colloid Interface Sci 1989, 30:153.
42. Block H, Rattray P: Recent developments in ER fluids. In Progress in Electrorheology. Edited by Havelka KO, Filisko FE. New York: Plenum Press; 1995:19-42.
43. See H: Advances in modeling the mechanisms and rheology of electrorheological fluids. Korea-Australia Rheol J 1999, 11:169.
44. Hao T: Electrorheological suspensions. Adv Colloid Interface Sci 2002, 97:1.
45. See H: Advances in electro-rheological fluids: Materials, modelling and applications. J Ind Eng Chem 2004, 10:1132.
46. Sung JH, Cho MS, Choi HJ, Jhon MS: Electrorheology of

semiconducting polymers. J Ind Eng Chem 2004, 10:1217.
47. Minagawa K, Koyama K: Electri- and magneto-rheological materials: stimuli-induced rheological functions. Curr Org Chem 2005, 9:1643.
48. Zhao XP, Yin JB: Advances in electrorheological fluids based on inorganic dielectric materials. J Ind Eng Chem 2006, 12:184.
49. Quadrat O, Stejskal J: Polyaniline in electrorheology. J Ind Eng Chem 2006, 12:352.
50. Zhao XP, Yin JB, Tang H: New advances in design and preparation of electrorheological materials and devices. In Smart Materials and Structures: New Research. Edited by Reece PL. New York: Nova Science Publishing; 2007:1-66.
51. Kim DH, Kim YD: Electrorheological properties of polypyrrole and its composite ER fluids. J Ind Eng Chem 2007, 13:879.
52. Choi HJ, Jhon MS: Electrorheology of polymers and nanocomposites. Soft Matter 2009, 5:1562.
53. Wen WJ, Huang XX, Yang SH, Lu KQ, Sheng P: The giant electrorheological effect in suspensions of nanoparticles. Nat Mater 2003, 2:727.
54. Lu KQ, Shen R, Wang XZ, Sun G, Wen WJ: The electrorheological fluids with high shear stress. Int J Mod Phys B 2005, 19:1065.
55. Cao GJ, Shen M, and Zhou LW: Electrorheological properties of triethanolamine modified amorphous TiO_2 electrorheological fluid. J Solid State Chem 2006, 179:1565.
56. Qiao YP, Yin JB, and Zhao XP: Oleophilicity and the strong electrorheological effect of surface-modified titanium oxide nano-particles. Smart Mater Struct 2007, 16:332.
57. Wen WJ, Huang XX, Sheng P: Particle size scaling of the giant electrorheological effect. Appl Phys Lett 2004, 85:299.
58. Wang BX, Zhao Y, Zhao XP: The wettability, size effect and electrorheological activity of modified titanium oxide nanoparticles. Colloids Surf A 2007, 295:27.

59. Gong XQ, Wu JB, Huang XX, Wen WJ, Sheng P: Influence of liquid phase on nanoparticle-based giant electrorheological fluid. Nanotechnology 2008, 19:165602.
60. Lu Y, Shen R, Wang XZ, Sun G, and Lu KQ: The synthesis and electrorheological effect of a strontium titanyl oxalate suspension. Smart Mater Struct 2009, 18:025012.
61. Cheng YC, Liu XH, Guo JJ, Liu FH, Li ZX, Xu GJ, Cui P: Fabrication of uniform core-shell structural calcium and titanium precipitation particles and enhanced electrorheological activities. Nanotechnology 2009, 20:055604.
62. Ahn KH, Klingenberg DJ: Relaxation of polydisperse electrorheological suspensions. J Rheol 1994, 38:713.
63. Kawai A, Uchida K, Ikazaki F: Effects of shape and size of dispersoid on electrorheology. In Proceedings of 8th International Conference: Electrorheological Fluids and Magnetorheological Suspensions. Edited by Bossis G. Singapore: World Scientific; 2002:626-632.
64. Yatsuzuka K, Miura K, Kuramoto N, Asano K: Observation of the electrorheological effect of silicone oil/polymer particles suspension. IEEE Trans IAS 1995, 31:457.
65. Asano K, Suto H, Yatsuzuka K: Influence of the particle configuration on electrorheological effect. J Electrostat 1997, 40-41:573.
66. Kanu RC, Shaw MT: Enhanced electrorheological fluids using anisotropic particles. J Rheol 1998, 42:657.
67. Otsubo Y: Electrorheology of whisker suspensions. Colloids Surf A 1999, 153:459.
68. Tsuda K, Takeda Y, Ogura H, Otsubo Y: Electrorheological behavior of whisker suspensions under oscillatory shear. Colloids Surf A 2007, 299:262.
69. Qi YB, Wen WJ: Influences of geometry of particles on electrorheological fluids. J Phys D Appl Phys 2002, 35:2231.
70. Kae YK, See H: The electrorheological response of elongated particles. Rheol Acta 2010, 49:741.

71. Feng P, Wan Q, Fu XQ, Wang TH: Anomalous electrorheological behavior of ZnO nanowires. Appl Phys Lett 2005, 87:033114.
72. Yin JB, Zhao XP: Titanate nano-whisker electrorheological fluid with high suspended stability and ER activity. Nanotechnology 2006, 17:192.
73. Yin JB, Zhao XP: Electrorheological properties of titanate nanotube suspensions. Colloids Surf A 2008, 329:153.
74. Kasuga T, Hiramatsu M, Hoson A, Sekino T, Niihara K: Formation of titanium oxide nanotube. Langmuir 1998, 14:3160.
75. He Y, Cheng Q, Pavlínek V, Li C, Sáha P: Electrorheological behavior of titanate nanotubes suspensions under oscillatory shear. J Phys Conf Ser 2009, 149:012010.
76. He Y, Cheng Q, Pavlínek V, Li C, Sáha P: Synthesis and electrorheological characteristics of titanate nanotube suspensions under oscillatory shear. J Ind Eng Chem 2009, 15:550.
77. Lozano K, Hernandez C, Petty TW, Sigman MB, Korgel B: Electrorheological analysis of nano laden suspensions. J Colloid Interface Sci 2006, 297:618.
78. Ramos-Tejada MM, Espin MJ, Perea R, Delgado AV: Electrorheology of suspensions of elongated goethite particles. J Non-Newtonian Fluid Mech 2009, 159:34.
79. Cheng YC, Wu KH, Liu FH, Guo JJ, Liu XH, Xu GJ, Cui P: Facile approach to large-scale synthesis of 1D calcium and titanium precipitate (CTP) with high electrorheological activity. ACS Appl Mater Interf 2010, 2:621.
80. Block H, Kelly JP: Electro-rheological fluids. 1986. UK Patent 217051B
81. Gow C, Zukoski CF: The electrorheological properties of polyaniline suspensions. J Colloid Interface Sci 1990, 136:175.
82. Xie HQ, Guan JG: Study on electrorheological properties of semiconducting polyaniline-based suspensions. Angew Makromol Chem 1996, 235:21.

83. Wu SZ, Lu S, Shen JR: Electrorheological suspensions. Polym Int 1996, 41:363.
84. Choi HJ, Kim TW, Cho MS, Kim SG, Jhon MS: Electrorheological characterization of polyaniline dispertions. Eur Polym J 1997, 33:699.
85. Quadrat O, Stejskal J, Kratochvíl P, Klason C, McQueen D, Kubát J, Sáha P: Electrical properties of polyaniline suspension. Synth Met 1998, 97:37.
86. Kuramoto N, Yamazaki M, Nagai K, Koyama K, Tanaka K, Yatsuzuka K, Higashiyama Y:Electrorheological property of a polyaniline-coated silica suspension. Thin Solid Films 1994, 239:169.
87. Gozdalik A, Wycislik H, Plocharski J: Electrorheological effect in suspensions of polyaniline. Synth Met 2000, 109:147.
88. Cho MS, Choi HJ, To K: Effect of ionic pendent groups on a polyaniline-based electrorheological fluid. Macromol Rapid Commun 1998, 19:271.
89. Lu J, Zhao XP: Electrorheological properties of a polyaniline montmorillonite clay nanocomposite suspension. J Mater Chem 2002, 12:2603.
90. Cho MS, Kim JW, Choi HJ, Jhon MS: Polyaniline and its modification for electroresponsive material under applied electric fields. Polym Adv Technol 2005, 6:352.
91. Akhavan J, Slack K: Coating of a semi-conducting polymer for use in electrorheological fluids. Polymer 2001, 124:363.
92. Chin BD, Park OO: Dispersion stability and electrorheological properties of polyaniline particle suspensions stabilized by poly (vinyl methyl ether). J Colloid Interface Sci 2001, 234:344.
93. Lu J, Zhao XP: A new approach of enhancing the shear stress of electrorheological fluids of montmorillonite nanocomposite by emulsion intercalation of poly-N-methaniline. J Colloid Interface Sci 2004, 273:651.
94. Park SJ, Park SY, Cho MS, Choi HJ, Jhon MS: Synthesis and electrorheology of multi-walled carbon nanotube/polyaniline

nanoparticles. Synth Met 2005, 152:337.
95. Fang FF, Lee BM, Choi HJ: Electrorheologically intelligent polyaniline and its composites. Macromol Res 2010, 18:99.
96. Yin JB, Zhao XP, Xia X, Xiang LQ, Qiao YP: Electrorheological fluids based on nano-fibrous polyaniline. Polymer 2008, 49:4413.
97. Liu YD, Fang FF, and Choi HJ: Core-shell structured semiconducting PMMA/Polyaniline snowman-like anisotropic microparticles and their electrorheology. Langmuir 2010, 26:12849.
98. Yin JB, Xia X, Xiang LQ, Qiao YP, Zhao XP: The electrorheological effect of polyaniline nanofiber, nanoparticle and microparticle suspensions. Smart Mater Struct 2009, 18:095007.
99. Xia X, Yin JB, Qiang PF, Zhao XP: Electrorheological properties of thermo-oxidative polypyrrole nanofibers. Polymer 2011, 52:786.
100. Kurachi Y, Toshiyuki O, Tanaka M, Ishino Y, Saito T: Electrorheological fluid. 1992. US Patent 5106521
101. Powell B: Preparation of electrorheological fluids using fullerenes and other crystals having fullerenes-like anisotropic electrical properties. 1995. US Patent 5445759.
102. Shima K, Yamamoto I, Senda Y: Electro-rheological fluid. 1994. JP Patent 06001990
103. Kojima Y, Tsuji M, and Matsuoka T, Takahashi H: Reaction of diphenyldiacetylene by annealing under elevated pressure. J Polym Sci Polym Chem 1994, 32:1371.
104. Kojima Y, Matsuoka T, Takahashi H: Electrorheological properties of suspension of diphenyldiacetylene annealed under elevated pressure. J Appl Polym Sci 1994, 53:1393.
105. Ishino Y, Maruyama T, Ohsaki T, and Endo S, Goshima N: Anhydrous electrorheological fluid using carbonaceous particulate as dispersed phase. In Progress in electrorheology. Edited by Havelka KO, Filisko FE. New York: Plenum Press; 1995:137-145.

106. Kojima Y, Matsuoka T, Takahashi H, Kurauchi T: Electrorheological properties of suspension of carbonaceous particles. J Mater Sci Lett 1995, 14:623.
107. Miyano M, Shi H, Saito T: Electro-rheological fluid. 1996. JP Patent 08143886
108. Sakurai R, See H, Saito T: The effect of blending particles with different conductivity on electrorheological properties. J Rheol 1996, 40:395.
109. Sasaki M, Kobayashi Y, Haji K: Electrorheological fluid containing carbonaceous particles. 1998. US Patent 5705088
110. Maruyama T, Ogino T, Ishino Y, Saito T, Haraoka T, Takagi K, Hatano H: Carbonaceous powder to be dispersed in electrorheological fluid and electrorheological fluid using the same. 1998. US Patent 5779880
111. Choi HJ, Kim JW, Yoon SH, Fujiura R, Komatsu M, Jhon MS: Synthesis and electrorheological characterization of carbonaceous particle suspensions. J Mater Sci Lett 1999, 18:1445.
112. Endo S, Wada H: Electrorheological fluid. 2002. US Patent, 6352651 B1
113. Negita K, Misono Y, Yamaguchi T, Shinagawa J: Dielectric and electrical properties of electrorheological carbon suspensions. J Colloid Interface Sci 2008, 321:452.
114. Dong P, Wang CH, and Zhao SQ: Preparation of high performance electrorheological fluids with coke-like particles from FCC slurry conversion. Fuel 2005, 84:685.
115. Schwarz M, Bauhofer W, Schulte K: Alternating electric field induced agglomeration of carbon black filled resins. Polymer 2002, 43:3079.
116. Bezryadin A, Westervelt RM, Tinkham M: Self-assembled chains of graphitized carbon nanoparticles. Appl Phys Lett 1999, 74:2699.
117. Svásand E, Helgesen G, Skjeltorp AT: Chain formation in a complex fluid containing carbon cones and disks in silicon oil. Colloids Surf A 2007, 308:67.

118. Qiao YP, Zhao XP: Electrorheological effect of carbonaceous materials with hierarchical porous structures. Colloids Surf A 2009, 340:33.
119. Sazonova V, Yaish Y, Ustunel H, Roundy D, Arias TA, McEuen PL: A tunable carbon nanotube electromechanical oscillator. Nature 2004, 431:284.
120. Jin HJ, Choi HJ, Yoon SH, Myung SJ, Shim SE: Carbon nanotube-adsorbed polystyrene and poly (methyl methacrylate) microspheres. Chem Mater 2005, 17:4034.
121. Park SJ, Park SY, Cho MS, Choi HJ, Jhon MS: Synthesis and electrorheology of multi-walled carbon nanotube/polyaniline nanoparticles. Synth Met 2005, 152:337.
122. Park SJ, Cho MS, Lim ST, Choi HJ, Jhon MS: Electrorheology of multiwalled carbon nanotube/poly (methyl methacrylate) nanocomposites. Macromol Rapid Commun 2005, 26:1563.
123. Lee IS, Yoon SH, Jin HJ, Choi HJ: Adsorption of multi-walled carbon nanotube onto poly (methyl methacrylate) microsphere and its electrorheology. Diam Relat Mater 2006, 15:1094.
124. Kim ST, Lim JY, Park BJ, Choi HJ: Dispersion-polymerized carbon nanotube/poly (methyl methacrylate) composite particles and their electrorheological characteristics. Macromol Chem Phys 2007, 208:514.
125. Han SJ, Kim B, Suh KB: Electrorheological properties of poly (acrylonitrile) microspheres coated with mutiwall carbon nanotubes. Mater Lett 2007, 61:3995.
126. Zhang K, Lim JY, Choi HJ, Seo Y: Core-shell structured carbon nanotube/poly (methyl methacrylate) composite and its electrorheological characteristics. Diam Relat Mater 2008, 17:1604.
127. Slobodian P, Pavlínek V, Lengálová A, Sáha P: Polystyrene/multi-wall carbon nanotube composites prepared by suspension polymerization and their electrorheological behavior. Curr Appl Phys 2009, 9:184.
128. Li MJ, Wang XB, Tian R, Liu FM, Hu HT, Chen R, Zheng H, Wan L: Preparation, solubility, and electrorheological

properties of carbon nanotubes/poly(methyl methacrylate) nanocomposites by in situ functionalization. Composites A 2009, 40:413.
129. Fang FF, Choi HJ, Jhon MS: Magnetorheology of soft magnetic carbonyl iron suspension with single-walled carbon nanotube additive and its yield stress scaling function. Colloids Surf A 2009, 351:46.
130. Li JX, Gong XQ, Chen SY, and Wen WJ, Sheng P: Giant electrorheological fluid comprising nanoparticles: Carbon nanotube composite. J Appl Phys 2010, 107:093507.
131. Brown MS, Shan JW, Lin C: Electrical polarizability of carbon nanotubes in liquid suspension. Appl Phys Lett 2007, 90:203108.
132. Lin C, Shan JW: Electrically tunable viscosity of dilute suspensions of carbon nanotubes. Phys Fluids 2007, 19:121702.
133. Lin C, Shan JW: Ensemble-averaged particle orientation and shear viscosity of single-wall-carbon-nanotube suspensions under shear and electric fields. Phys Fluids 2010, 22:022001.
134. Yin JB, Xia X, Xiang LQ, Zhao XP: Conductivity and polarization of carbonaceous nanotubes derived from polyaniline nanotubes and their electrorheology when dispersed in silicone oil. Carbon 2010, 48:2958.
135. Yin JB, Xia X, Xiang LQ, Zhao XP: Temperature effect of electrorheological fluids based on polyaniline derived carbonaceous nanotubes. Smart Mater Struct 2011, 20:015002.
136. Takahashi T, Murayama T, Higuchi A, Awano H, Yonetake K: Aligning vapor-grown carbon fibers in polydimethylsiloxane using dc electric or magnetic field. Carbon 2006, 44:1180.
137. Konishi M, Nagashima T, Asako Y: ER properties of suspension of polymer graft carbon black particles. In Proceedings of 6th International Conference on ERF and MRS. Edited by Nakano M, Koyama K. Singapore: World Scientific; 1998:12-17.

138. Inoue A: Study of a new electrorheological fluid. In Proceeeing of 2nd International Conference on Electrorheological Fluids. Edited by Carlson JD, Sprecher AF, and Conrad H. Lancaster-Basel: Technomic; 1990:176-183.
139. Tam WY, Yi GH, Wen W, Ma H, Loy MMT, Sheng P: New electrorheological fluid: theory and experiment. Phys Rev Lett 1997, 78:2987.
140. Saito T, Anzai H, Kurada S, Osawa Z: Electrorheological properties composed of polymer core with controlled diameter and electro-conductive/nonconductive double layers shell. In Proceedings of 6th International Conference on ERF and MRS. Edited by Nakano M, Koyama K. Singapore: World Scientific; 1998:19-24.
141. Luo CR, Tang H, Zhao XP: Design and preparation of a kind of metal/P (MMA-MAA) ERF. Int J Mod Phys B 2001, 15:672.
142. Cho MS, Cho YH, Choi HJ, Jhon MS: Synthesis and electrorheological characteristics of polyaniline-coated poly (methyl methacrylate) microsphere: Size effect. Langmuir 2003, 19:5875.
143. Davis L: The metal-particle/insulating oil system: An ideal electrorheological fluid. J Appl Phys 1993, 73:680.
144. Tang X, Conrad H: The role of dielectric relaxation in transient electrorheological response. J Appl Phys 1996, 80:5240.
145. Wu CW, Conrad H: Influence of a surface film on the particles on the electrorheological response. J Appl Phys 1997, 81:383.
146. Wu CW, Conrad H: Influence of a surface film on conducting particles on the electrorheological response with alternating current fields. J Appl Phys 1997, 81:8057.
147. Wu CW, Conrad H: Multi-coated spheres: recommended electrorheological particles. J Phys D Appl Phys 1998, 31:3312.
148. Yin JB, Xia X, Xiang LQ, Zhao XP: Coaxial cable-like polyaniline@titania nanofibers: facile synthesis and low power electrorheological fluid application. J Mater Chem 2010, 20:7096.

149. Liu YD, Fang FF, and Choi HJ: Silica nanoparticle decorated conducting polyaniline fibers and their electrorheology. Mater Lett 2010, 64:154.
150. Cheng Q, Pavlinek V, He Y, Li C, Saha P: Electrorheological characteristics of polyaniline/titanate composite nanotube suspensions. Colloid Polym Sci 2009, 287:435.
151. Luo M, He Y, Cheng Q, Li C: Synthesis and structural and electrical characteristics of polypyrrole nanotube/TiO_2 hybrid composites. J Macromol Sci B Phys 2010, 49:419.
152. Chuangchote S, Sirivat A, Supaphol P: Mechanical and electrorheological properties of electrospun poly (vinylalcohol) nanofibre mats filled withcarbon black nanoparticles. Nanotechnology 2007, 18:145705.
153. Pu HT, Jiang FJ: Towards high sedimentation stability: magnetorheological fluids based on CNT/Fe_3O_4 nanocomposites. Nanotechnology 2005, 16:1486.
154. López-López MT, Vertelov G, Bossis G, Kuzhir P, Duran JDG: New magnetorheological fluids based on magnetic fibers. J Mater Chem 2007, 17:3839.
155. Bell RC, Miller ED, Karli JO, Vavreck AN, Zimmerman DT: Influence of particle shape on the properties of magnetorheological fluids. Int J Mod Phys B 2007, 21:5018.
156. Bell RC, Karli JO, Vavreck AN, Zimmerman DT, Ngatu GT, Wereley NM: Magnetorheology of submicron diameter iron microwires dispersed in silicone oil. Smart Mater Struct 2008, 17:015028.
157. Ngatu GT, Wereley NM, Karli JO, and Bell RC: Dimorphic magnetorheological fluids: exploiting partial substitution of microspheres by nanowires. Smart Mater Struct 2008, 17:045022.
158. Zimmerman DT, Bell RC, Filer JA, Karli JO, Wereley NM: Elastic percolation transition in nanowire-based magnetorheological fluids. Appl Phys Lett 2009, 95:014102.
159. López-López MT, Kuzhir P, Bossis G: Magnetorheology of fiber suspensions. I. Experimental. J Rheol 2009, 53:115.

160. Kuzhir P, López-López MT, Bossis G: Magnetorheology of fiber suspensions. II. Theory. J Rheol 2009, 53:127.
161. de Vicente J, Segovia-Gutiérrez JP, Andablo-Reyes E, Vereda F, Hidalgo-Álvarez R:Dynamic rheology of sphere- and rod-based magnetorheological fluids. J Chem Phys 2009, 131:194902.
162. Gomez-Ramirez A, López-López MT, Durán JDG, and Gonzalez-Caballero F: Influence of particle shape on the magnetic and magnetorheological properties of nanoparticle suspensions. Soft Matter 2009, 5:3888.
163. Badescu V, Udrea1 LE, Rotariu1 O, Badescu R, Apreotesei G: The Cotton-Mouton effect in ferrofluids containing rod-like magnetite particles. J Phys Conf Ser 2009, 149:012101.

Chapter 2

Growth of Vertically Aligned ZnO Nanorods Using Textured ZnO Films

Francisco Solís-Pomar[1,2], Eduardo Martínez[3], Manuel F Meléndrez[4], and Eduardo Pérez-Tijerina[1,2]

[1]Centro de Innovación, Investigación y Desarrollo en Ingeniería y Tecnología de la UANL-PIIT, Apodaca, Nuevo León 66600, México

[2]Facultad de Ciencias Físico-Matemáticas, Universidad Autónoma de Nuevo León, San Nicolás de los Garza, Nuevo León 66451, México

[3]Centro de Investigación en Materiales Avanzados S. C., Unidad Monterrey-PIIT, Apodaca, Nuevo León 66600, México

[4]Department of Materials Engineering, (DIMAT), Faculty of Engineering, University of Concepción, 270 Edmundo Larenas, Casilla 160-C, Concepción, Chile

ABSTRACT

A hydrothermal method to grow vertical-aligned ZnO nanorod arrays on ZnO films obtained by atomic layer deposition (ALD) is presented. The growth of ZnO nanorods is studied as function of the crystallographic orientation of the ZnO films deposited on silicon (100) substrates. Different thicknesses of ZnO films around 40 to 180 nm were obtained and characterized before carrying out the growth process by hydrothermal methods. A textured ZnO layer with preferential direction in the normal c-axes is formed on substrates by the decomposition of diethylzinc to provide nucleation sites for vertical nanorod growth. Crystallographic orientation of the ZnO nanorods and ZnO-ALD films was determined by X-ray diffraction analysis. Composition, morphologies, length, size, and diameter of the nanorods were studied using a scanning electron microscope and energy dispersed x-ray spectroscopy analyses. In this work, it is demonstrated that crystallinity of the ZnO-ALD films plays an important role in the vertical-aligned ZnO nanorod growth. The nanorod arrays synthesized in solution had a diameter, length, density, and orientation desirable for a potential application as photosensitive materials in the manufacture of semiconductor-polymer solar cells.

BACKGROUND

ZnO wurtzite hexagonal phase is one of the most important functional materials due to its excellent physicochemical properties and its diversity in terms of morphologies, properties, and applications[1,2]. The excellent properties of ZnO include direct band gap (3.37 eV) and high optical gain of 300 cm^{-1} (100 cm^{-1} for GaN) at room temperature, large saturation velocity (3.2×10^7 cm/s), high breakdown voltage, and large exciton binding energy (60 meV). These versatile properties of ZnO provide an opportunity to recognize it as one of the most multifunctional materials; therefore it can be used for ultraviolet lasers, light-emitting diodes, photo-detectors, piezoelectric transducers and actuators, hydrogen storage, chemical or biosensors, surface acoustic-wave guides, solar cells, and photo catalysts, among others [2-4]. As mentioned, one of the qualities of this material is that it can be obtained in different types of nanostructures, being 1D ZnO nanostructures such as nanorods and nanowires the most used owing to their great prospects in fundamental physical science, nanotechnology applications, nano-optoelectronics, and photovoltaic devices. Hence, it is desirable to develop fast, simple, and mild routes for the synthesis of 1D high crystalline quality ZnO nanostructures in a large area, to explore their diverse applications.

Among various applications of this material, one can say that the utilization of ZnO nanostructures as photo-electrodes in dye-sensitized solar cells (DSSCs) has received considerable attention currently due to their compatibility with the commonly used TiO_2 materials [5-8]. Besides, ZnO shows higher electronic mobility and similar energy level of the conduction band than TiO_2 which makes ZnO a candidate to be used as a photo-electrode material for the fabrication of efficient DSSCs. Several methods have been used to grow nanowires and nanorods such as: vapor-liquid-solid (VLS) [3,4], metal organic vapor-phase epitaxy (MOVPE) [9], pulsed laser deposition (PLD) [5,6] solution, and hydrothermal methods [7,8].

In some cases, these arrays were synthesized at temperatures ranging from 400°C to 600°C through metal-organic chemical vapor deposition (MOCVD) [10-12], PLD [13], and chemical vapor transport (CVT) [14] implying high temperature, complex, and expensive processes. In addition, high-quality vertical ZnO nanowire arrays have been grown using both (1) heteroepitaxy with Al_2O_3 or single-crystalline GaN, which is currently limited to expensive substrates [15-17] and (2) homoepitaxy with a textured ZnO thin film that is deposited on top of a non-epitaxial substrate which act as a nanorod nucleation layer [18-22]. Neither approach is particularly low-cost, versatile, or promising for the fabrication of high-performance ZnO nanowire optoelectronic devices, including solar cells.

With the aim to explain the nanorod alignment, Zhang et al. hypothesized that a textured ZnO wetting layer formed prior to nanorod growth favors the alignment [22]. If this notion is correct, it should be possible to control the crystallography of the seed layer film to obtain nanorods by an alternative method like the hydrothermal treatment and thus enhance the process to achieve aligned nanorods. Therefore, one could then create surfaces that would work as growth seeds for the ZnO nanowires on an assortment of substrates using any nanowire growth technique, e.g., gas-phase or solution phase. In this work, vertically aligned ZnO nanorods on Si(100) substrates were synthesized using a hybrid atomic layer deposition (ALD) and hydrothermal method. For accomplishing this, ZnO films were prepared by ALD at different thicknesses to obtain seed layers of different crystallographic nature.

The purpose of this work is to study the effect of crystalline orientation of the seed layer on the ZnO nanorods growing by hydrothermal. In this way the aim is to determine the best conditions to grow perfectly aligned and uniform ZnO nanorods and provide the foundation to achieve a better controlled and large-scale synthesis of ZnO nanorods.

EXPERIMENTAL

The fabrication procedure for the growth of the nanorods consists of two steps: (1) preparation of a seed-textured ZnO thin layer by ALD and (2) the nanorod array growth by hydrothermal.

Synthesis of ZNO Films by ALD

ZnO films with different thicknesses, 40, 80, 120, and 180 nm were deposited on Si(100) substrates by ALD using a Savannah 100 ALD system from Cambridge Nanotech. Diethylzinc (DEZn) was used as the precursor for zinc and deionized water was used as the oxidation source. The growth cycle consists of precursor exposures and N_2 (99.9999%) purge following the sequence of DEZn/N_2/H_2O/N_2 with corresponding duration of 0.1:5:0.1:5 s. After each N_2 purging, the reactor was pumped down to 0.1 Torr. DEZn and H_2O

were fed into the chamber through separate inlet lines and nozzles. In the ALD method, reagents (precursors) are introduced sequentially into the growth chamber and when precursors reach the substrate, they are interspersed by cycles of purging with inert gas (N_2). The opening and closing sequences of the valves were controlled by a computer. Precursor introduction was done by opening the inlet valve between the reservoir and reactor chamber while the outlet valve was closed. The pressures of the DEZn and H_2O in the reactor chamber were approximately 1 and 2 Torr, respectively, monitored by a vacuum gauge. The substrate temperature was maintained at 177°C during the deposition. The reaction was repeated 400, 800, 1,200, and 1,800 cycles to obtain the ZnO films with different thicknesses and crystallographic features.

Growth of ZNO Nanorods Through Hydrothermal Process

In this process, $Zn(NO_3)_2$ (ZNT) and hexamethylenetetramine (HMT) purchased from Sigma-Aldrich (St. Louis, MO, USA) were used as reagents. The ZnO nanorods were grown in aqueous solutions of zinc nitrate ($Zn(NO_3)_2.6H_2O$) 0.01 M and hexamine ($(CH_2)_6N_4$) in deionized water; the ZNT/HMT molar ratio was always {1:1}. The ALD-ZnO films were placed in face-up position into glass reactor with screw cap and then equal amounts of both ZNT and HMT solutions were added.

The reactor was immersed in a water bath at 90°C with mild agitation during 4 h. Finally the samples were rinsed with deionized water for several times and dried at 90°C for several hours before characterization. The samples were structurally and morphologically characterized by X-ray diffraction (XRD) using a Philips X'Pert PW3040 diffractometer (PANalytical, Almelo, the Netherlands) with Cu-K radiation and field emission scanning electron microscopy in a Hitachi S-5500 Field Emission Gun (Hitachi Co., Tokyo, Japan) ultrahigh-resolution scanning electron microscope (FE-SEM) (0.4 nm at 30 kV) with a BF/DF Duo-STEM detector. Additionally, the composition was determined by energy dispersive X-ray spectroscopy (EDS) with an INCA-Energy EDS (Oxford Instruments, Oxfordshire, UK) attached to the FE-SEM; and the seed-textured ZnO layer surface was analyzed by atomic force microscopy (AFM).

RESULTS AND DISCUSSION

Zno Films by ALD

XRD was performed on both substrates before and after nanorod growth. The crystallinity of the grown ZnO films obtained by ALD is shown in a typical XRD pattern in Figure 1. The X-ray spectra show well-defined Bragg peaks for the ZnO films corresponding to the planes (100), (002), and (110); these also confirm the wurtzite crystal structure of the whole set of samples (wurtzite structure, $a = 3.249$ Å and $c = 5.201$ Å) which is consistent with data of ZnO JCPDS no. 36-1451. All films were polycrystalline and at room temperature the strong signal centered at 34.5 indicates preferential growth in the (002) direction because the c-plane perpendicular to a substrate is the most densely packed and thermodynamically favorable plane in the wurtzite structure. This crystallographic condition induces some kind of c-axis texturing which depends of thickness. The degree of the orientation as function of thickness can be illustrated by the relative texture coefficient, which is given by Eq. 1:

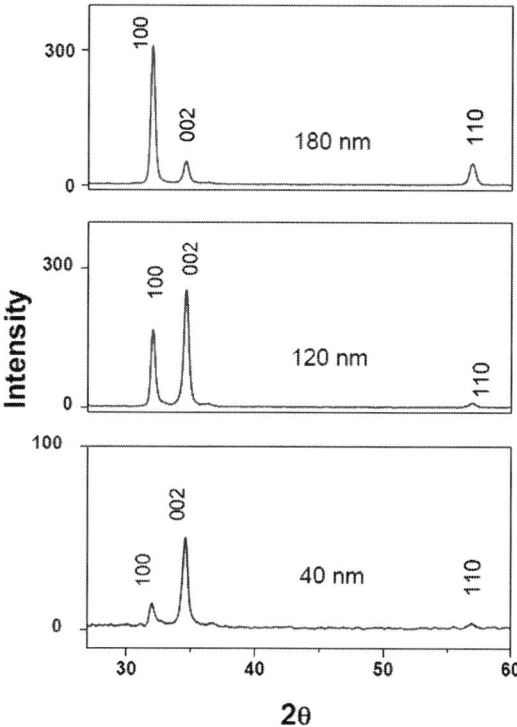

Figure 1: X-ray patterns of ZnO films. ZnO films with thicknesses between 40 and 180 nm.

$$TC_{002} = (I_{002}/I_{002}^0)/(I_{002}/I_{002}^0 + I_{100}/I_{100}^0) \qquad (1)$$

where TC_{002} is the relative texture coefficient of diffraction peaks (002) over (100), I_{002} and I_{100} are the measured diffraction intensities due to (002) and (100) planes, respectively, and I_{002}^0 and I_{100}^0 are the corresponding values of standard PDF(36- 1451) measured from randomly oriented powder samples, so on this basis one can say that for materials with random crystallographic orientations, e.g., powders, the texture coefficient is 0.5. Now, about those ALD-ZnO films in which the highest peak was (002), as occurs in 40 and 120 nm films, the corresponding TC_{002} was increased as a confirming evidence of a preferential growth in that direction. The texture coefficient was 0.81, 0.60, and 0.14 for 40, 120, and

180 nm, respectively. It is also observed that preferential growth is disrupted with the increase of thickness given that the (100) peak at 31.7 becomes more intense for 180-nm films; it has been considered that the < 100 > orientation is favored due to the atomic disorder promoted with the ALD growth time. Texturing is apparently dependent of growth time because at longer times a crystallographic disorder is developed which limit the c-axis-oriented seeds and the crystal domain size. High texture in < 001 > direction will determine the quality of alignment and seed size the diameter of nanorods.

AFM images of ALD-ZnO films grown with different thicknesses are shown in Figure 2 to distinguish typical surface features previous to the hydrothermal process. These micrographs depict that with the thickness increasing, their roughness and surface defects also increase, thus allowing the formation of nucleation sites for ZnO nanorods growth. The ZnO films are composed of fine small grains (seeds) and these have average height (AH) that depends on the film thickness, if the ALD-ZnO films of 40 and 120 nm are observed one can see AH values of 18.2 and 31.4 nm, respectively. The differences in crystallographic and microstructural properties are significantly influenced by the ALD parameters such as the process time and flow rate.

The increase of roughness could influence the ZnO nanorod growth due to the fact that surface defects augment acting as a barrier to nucleation sites. It must be a competence between the number of nucleation sites and the crystallographic orientation disrupted by surface defects formed at the ALD-ZnO film. Table 1 shows the measurements developed through the AFM images as shown in Figure 2; here, it is evident that a long-term ALD deposit leads to create higher surface defects that must have an influence for the nanorod growth as it is demonstrated by scanning electron microscopy (SEM) analysis. For films with thickness of 40, 80, 120, and 180 nm the roughness was 3.2, 5.5, 8.1, and 12 nm, respectively. From these results, it is evident that surface roughness is greater when the film thickness increases. Maximums at the surface are high-energy sites where nanorod nucleation will be privileged while depression sites could be the non-growth regions due to the absence of oriented seeds that favors ZnO nanorod growth.

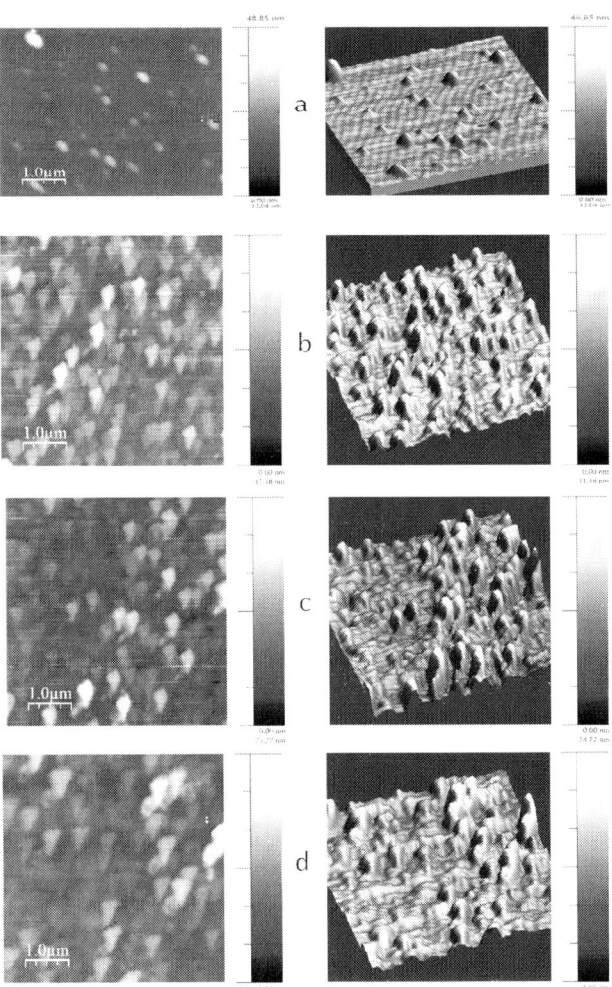

Figure 2: AFM images ZnO films with different thicknesses**:** **(a)** 40 nm,**(b)** 80 nm, **(c)** 120 nm, and **(d)** 180 nm.

Table 1: AFM features (roughness and height of textured ALD-ZnO films)

Cycles	Mean roughness (nm)	Mean height (nm)
400	3.2	18.18
800	5.5	19.25
1,200	8.1	20.73
1,800	12	31.41

After the nanorod growth on ALD-ZnO films with different textures, X-ray spectra were also recorded as depicted in Figure 3. XRD patterns of the resulting nanorod growth demonstrate that the orientation of the seed-textured ZnO films directly determines the orientation of the nanorods grown on these films.

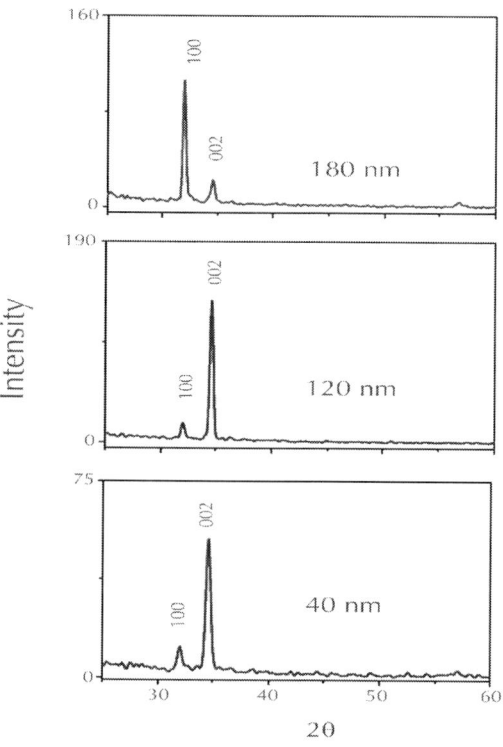

Figure 3: **X-ray patterns of ZnO nanorods.** ZnO nanorods grown on ALD-ZnO films with thicknesses between 40 and 180 nm.

From spectra, it is evident that the order of importance in intensity is maintained but the intensity ratio is changed as function of the nanorods growth type. In those ALD-ZnO films, in which the highest peak was (002) as occurs in 40, 80, and 120-nm films, the texture coefficient TC_{002} was increased as a confirming evidence of a preferential growth in that direction. The results indicate that the ZnO nanorod arrays are highly aligned on Si(100) substrate with c-axial growth direction, in addition, the diffraction intensity of the (002) peak surpasses others, which illustrates the c-oriented nature of the grown array. Otherwise, the TC_{002} of samples grown on textured ZnO films for 40, 120, and 180 nm is 0.84, 0.9, and 0.16, respectively; therefore the XRD results suggest that our samples are wurtzite ZnO nanorods with preferential c-orientation as confirmed by SEM analysis.

Figure 4 shows SEM images of the ZnO nanorod array grown by hydrothermal process on ALD-ZnO films with different thicknesses. The SEM images show a top view of the material deposited on the seed layer. It can be seen that density of ZnO nanorods depends on film thickness, whereas low density is typical from 40-nm films in Figure 4a, high density is present when a 120-nm film is used as seed layer in Figure 4b. Apparently thicknesses below 120 nm related with short ALD deposits give seeded surfaces with highly c-axis-oriented seeds whose size determines the nanorod diameter. Small thickness leads to small seed domain and thus, small diameters and low density of nanorods while long-term ALD experiments disrupt the ordered growth. The best conditions occur for middle-term ALD deposits in which the c-axis orientation is preserved and size domain increases to get larger diameters. The length of nanorods seems to be more dependent of hydrothermal process duration. The SEM images were also recorded in cross-section view to determine length and thickness for nanorods as shown in Figure 5. The measure of the nanorods size, population, thickness, and length was randomly chosen and obtained data were represented to obtain mean values. Therefore, the average nanorod size was fitted.

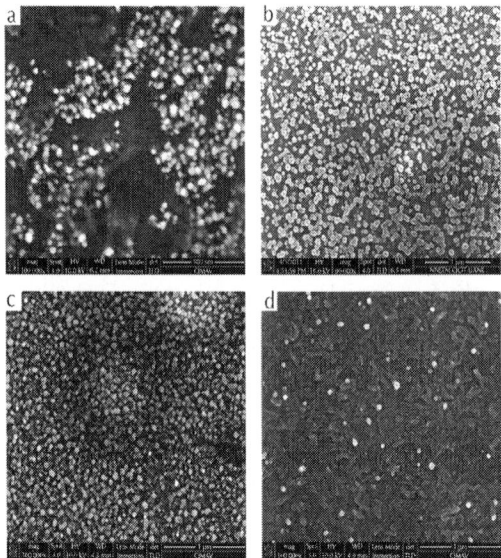

Figure 4: **Top view SEM images**. Images of ZnO nanorods grown on ALD-ZnO films: **(a)** 40 nm, **(b)** 80 nm, **(c)** 120 nm, and **(d)** 180 nm.

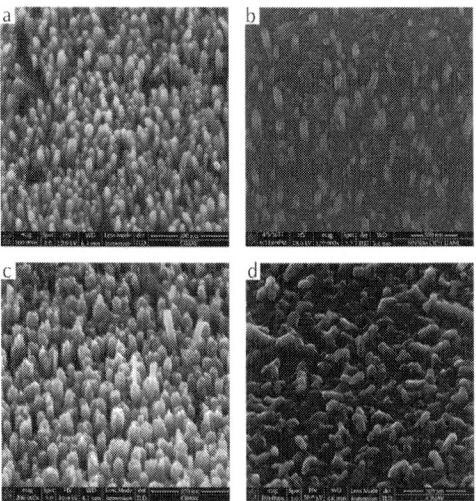

Figure 5: Tilted SEM images. Tilted images of ZnO nanorods grown on ALD-ZnO films of **(a)** 40 nm, **(b)** 80 nm, **(c)** 120 nm, and **(d)** 180 nm grown at 90°C, 4 h.

The nanorods have a narrow size distribution centered at about 34.5 ± 3.9 nm in diameter for the 40-nm films and 51.5 ± 5.2 nm for the 120-nm films. Cross-section view in Figure 5 demonstrated that the ZnO nanorods grew vertically with a mean length about 75.7 ± 14.3 nm for the 40 nm-films and 344.1 ± 97.6 nm for the 120-nm films. These geometric parameters are tunable to varying degrees by changing the growth time, ZNT concentration, or crystallography of seed-textured films. These results implied that our method is applicable to mass production of well-aligned ZnO nanorod arrays. All these results confirm that the hybrid method proposed to support nanorods is effective due to their high uniform distribution far and wide of the conducting substrate surface. The combined XRD and SEM data strongly suggest that c-axis texturing occurs across the ALD-ZnO film. A tilted SEM image of ZnO nanorods grown on an 80-nm ALD-ZnO film is presented in Figure 6a to confirm that nanorod growth also occurs at thicknesses within the 40 to 120 nm range. On the other hand, Figure 6b shows the chemical composition of the nanorods determined by EDS. Only oxygen, zinc, and silicon are detected to confirm that the ZnO nanorods are the only phase present.

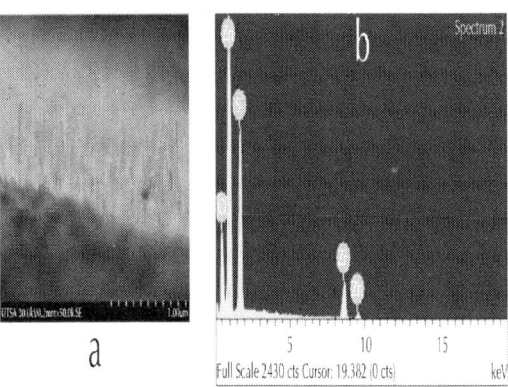

Figure 6: **Tilted SEM image and EDS spectra.** (a) Tilted image for ZnO nanorods grown on ALD-ZnO films of 80 nm and (b) EDS spectra to state the chemical nature of grown nanorods.

Chemical Reaction and Growth Mechanism

As stated by other authors, it is considered that the following reactions are involved in the crystal growth of ZnO nanorods [23-28].

$$C_6H_{12}N_4 + 6H_2O \leftrightarrow 6CH_2O + 4NH_3 \tag{2}$$

$$(CH_2)_6N_4 + Zn^{2+} \rightarrow [Zn(CH_2)_6N_4]^{2+} \tag{3}$$

$$NH_3 + H_2O \leftrightarrow NH_4^+ + OH^- \tag{4}$$

$$Zn^{2+} + 4NH_3 \rightarrow Zn(NH_3)_4^{2+} \tag{5}$$

$$Zn^{2+} + 4OH^- \rightarrow Zn(OH)_4^{2-} \tag{6}$$

$$Zn(NH_3)_4^{2+} + 2OH^- \rightarrow ZnO + 4NH_3 + H_2O \tag{7}$$

$$Zn(OH)_4^{2-} \rightarrow ZnO + H_2O + 2OH^- \tag{8}$$

$$[Zn(CH_2)_6N_4]^{2+} + 2OH^- \rightarrow ZnO + H_2O + (CH_2)_6 \tag{9}$$

$(CH_2)_6N_4$ is disintegrated into formaldehyde (CH_2O) and ammonia (NH_3) as shown in Eq. 2. Ammonia tends to disintegrate water to produce OH⁻ anions as described in Eq. 4. Finally, OH⁻ anions react with zinc (II) cations to form $Zn(OH)_4^{2-}$ (Eq. 6). In the growth process of ZnO nanorods, the concentration of OH⁻ anions is the dominant factor. Therefore, $(CH_2)_6N_4$ that supplies OH⁻ anions plays an important key role in the growth of ZnO nanorods. Under the given pH and temperature, zinc (II) is thought to exist primarily as $Zn(NH_3)_4^{2+}$ and $Zn(OH)_4^{2-}$. The ZnO is formed by the dehydration of these intermediates. The solution method used a closed system that contains limited amounts of precursor. Along with the heterogeneous nanorod growth on the ZnO seed layer, there is also homogeneous nucleation of ZnO crystals in solution. This homogeneous nucleation consumes ZnO precursors rapidly and causes early termination of growth on the substrate. Therefore, depletion of the precursor is inevitable and growth rate decreases as reaction time increases.

The reason for the c-axis-aligned nanorods is now examined. The microscopic details of seed formation have not been sufficiently understood and clarified to pinpoint which mechanism is responsible for the nanorod alignment. Some facts related with mechanisms at high temperatures, electrostatic processes, and electrical stability achieved by an exchange of charge mediated by surface states have been recently reported [26]. However, an explanation can be proposed in terms of our textured ALD-ZnO films. Textured ZnO films provide a surface formed mainly by seeds with c-axis-preferred orientation; these exposed basal planes of hexagonal rods are polar and have relatively high surface energy. As a result, the polar top planes attract more ion species promoting a faster growth rate and with this, the vertical-aligned ZnO nanorods emerge from the substrate. With all the mentioned before, it is reasonable to expect that ZnO nanorods orientation is determined by the nucleation and growth of the first few layers of zinc and oxygen atoms at the ALD seed layer through the fastest growth direction. This occurs because the polar {001} faces of wurtzite ZnO are electrostatically unstable and cannot exist without a mechanism to redistribute their surface charge and lower their free energy. According to reported models, optimized {001} surfaces have roughly 60% higher cleavage energy than the nonpolar {100} and {110} faces. Polar surfaces are generally stabilized by surface reconstruction or faceting; transfer of charge between surfaces or surface nonstoichiometry, including the neutralization of surface charge by adsorbed molecules. The following could enable the c-axis aligned nanorods: (1) Molecules present under the hydrothermal conditions adsorb onto nascent {001} surfaces and stabilize them relative to competing facets. In the decomposition of zinc nitrate to ZnO, these adsorbates would be primarily hydroxyl groups. The growth is favored due to the preference space of the reacting species, as illustrated in Figures 7 and 8. This shows the structure for the face (001), dots above of the polyhedral structure correspond to OH surface groups. The growth process is facilitated by the tetrahedral structure of the species $Zn[(OH)_4]^{2-}$ which fits well with the (001) polyhedral surface, this spatial resonance increases the growth in this direction more that in another faces. (2) The {001} surface energy depends on the crystal thickness so that very thin ZnO crystals prefer a {001} orientation, which is then kinetically locked-in as growth proceeds. (3) The first few atomic layers of ZnO must adopt a low-energy configuration different from the bulk lattice and later convert to the (001) orientation by a minor structural transformation. Notwithstanding all mentioned above, it is deemed that microscopic analysis of seed formation must be developed to pinpoint the right mechanism responsible for nanorod alignment.

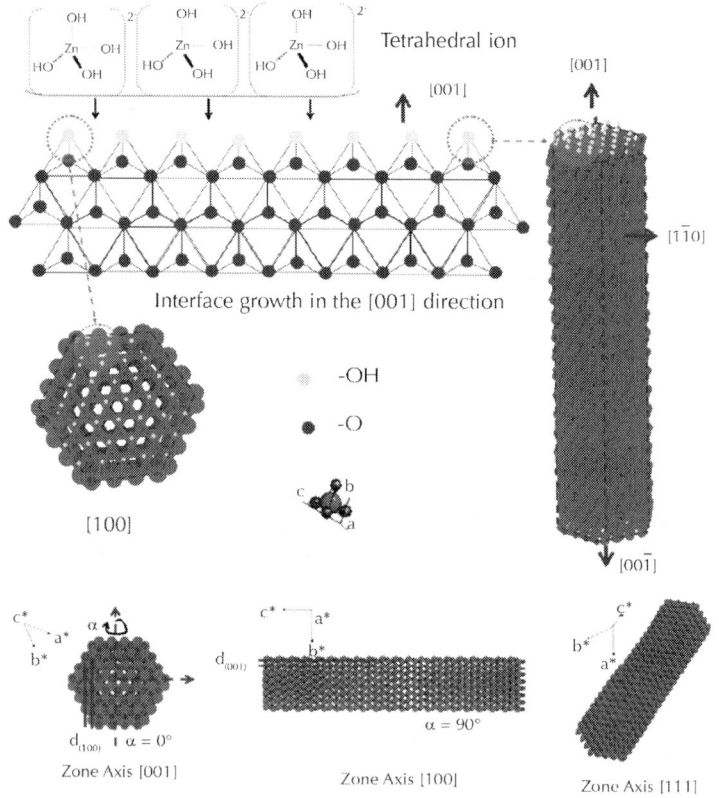

Figure 7: Growth mechanism. Proposed mechanism for ZnO nanorods growth at [001] direction.

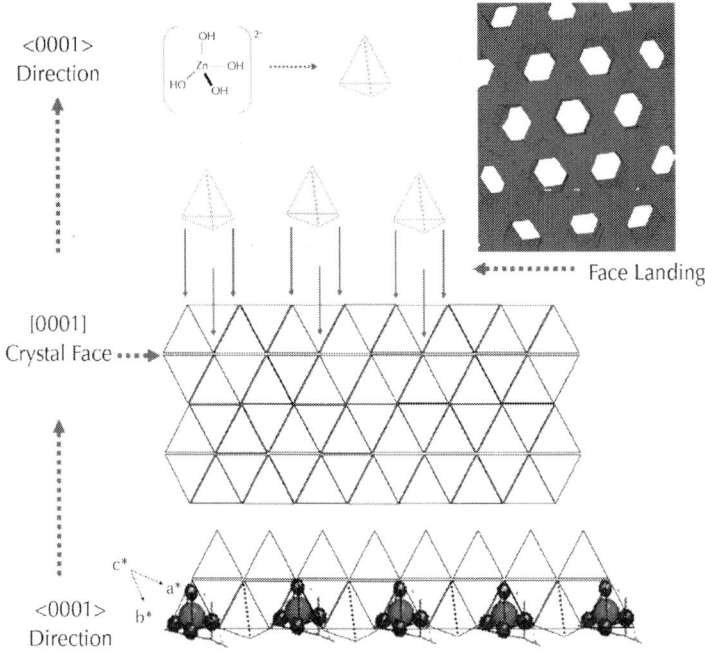

Figure 8: Growth process of ZnO nanorods in the direction [001]. The growth process is facilitated by the tetrahedral structure of the species $Zn[(OH)_4]^{2-}$ which fits well with the (001) surface polyhedra, this phenomenon (spatial resonance) increases the growth in this direction more than in another faces.

CONCLUSIONS

A simple seeding method for producing vertical ZnO nanorod arrays on Si(100) substrates is presented. By forming layers of textured ZnO films by ALD on a substrate, a seeded surface can be used to fabricate high-density vertical nanorod arrays. From the results, it is observed that thickness influences the texture of ALD-ZnO layer and thus, the crystallographic nature of the seed layer that determines the ulterior nanostructure growth type. Whereas short-term ALD deposit leads to create a surface with mostly c-axis-oriented seeds that favor the alignment, a long-term ALD deposit leads to create higher surface defects with polycrystalline seeds that promote disorder for ZnO nanorod growth. It is known that geometric parameters are tunable by changing the growth time and solution composition, with regards to the results of this work; this is also possible through changing seed density by controlling texture. Small thicknesses related with a short ALD deposit give seeded surfaces with small highly c-axis-oriented domains that promote small diameters with low density of nanorods while long-term ALD experiments disrupt the ordered growth. The best conditions occur for middle-term ALD deposits in which the c-axis orientation is preserved and size domain increases to get bigger diameters for nanorods. The arrays grown from aqueous solution feature a nanorod diameter, length, density, and orientation that make them highly suitable as the inorganic scaffold in efficient nanorod-polymer solar cells.

AUTHORS' CONTRIBUTIONS

FP carried out the hydrothermal synthesis and drafted the manuscript. EM carried out the ALD-ZnO textured substrates, developed the XRD, AFM, and SEM studies and drafted the manuscript. MFM participated in discussion of results contributing with his experience on this topic and contributed with the writing of manuscript. EP contributed with fruitful discussions to the presented research. All authors read and approved the final manuscript.

ACKNOWLEDGMENTS

For technical assistance in structural analysis, A. Toxqui, J. Aguilar, and N. Pineda are acknowledged. The authors are also grateful for the financial support of CONACyT through basic science projects 133252 and 118882.

REFERENCES

1. Li Z, Xiong Y, Xie Y: Selected-control synthesis of ZnO nanowires and nanorods via a PEG-assisted route.Inorg Chem 2003, 42:8105-8109.
2. Wang J, An X, Li Q, Egerton RF: Size-dependent electronic structures of ZnO nanowires.Appl Phys Lett 2005, 86:201911-201913.
3. Yang L, Yang J, Wang D, Zhang Y, Wang Y, Liu H, Fan H, Lang J: Photoluminescence and Raman analysis of ZnO nanowires deposited on Si(1 0 0) via vapor-liquid-solid process. Physica E: Low-dimensional Systems and Nanostructures 2008, 40:920-923.Shafiei S, Nourbakhsh A, Ganjipour B, Zahedifar M, Vakili-Nezhaad G: Diameter optimization of VLS-synthesized ZnO nanowires, using statistical design of experiment.Nanotechnology 2007, 18:355708.Son HJ, Jeon KA, Kim CE, Kim JH, Yoo KH, Lee SY: Synthesis of ZnO nanowires by pulsed laser deposition in furnace.Appl Surf Sci 2007, 253:7848-7850.Tien L, Pearton S, Norton D, Ren F: Synthesis and microstructure of vertically aligned ZnO nanowires grown by high-pressure-assisted pulsed-laser deposition.J Mater Sci 2008, 43:6925-6932.Wen B, Huang Y, Boland JJ: Controllable growth of ZnO nanostructures by a simple solvothermal process.J Phys Chem C 2008, 112:106-111.Guo M, Diao P, Cai S: Hydrothermal growth of well-aligned ZnO nanorod arrays: Dependence of morphology and alignment ordering upon preparing conditions.
4. J Solid State Chem 2005, 178:1864-1873.Robin IC, Gauron B, Ferret P, Tavares C, Feuillet G, Dang LS, Gayral B, Gerard JM:Evidence for low density of nonradiative defects in ZnO nanowires grown by metal organic vapor-phase epitaxy.Appl Phys Lett 2007, 91:143120-143122.Park WI, Yi GC, Kim M, Pennycook SJ: ZnO nanoneedles grown vertically on Si substrates by non-catalytic vapor-phase epitaxy.Adv Mater 2002, 14:1841-1843.

5. Wu JJ, Liu SC: Low-temperature growth of well-aligned ZnO nanorods by chemical vapor deposition.Adv Mater 2002, 14:215-218.Yuan H, Zhang Y: Preparation of well-aligned ZnO whiskers on glass substrate by atmospheric MOCVD.J Cryst Growth 2004, 263:119-124.Sun Y, Fuge GM, Ashfold MNR: Growth of aligned ZnO nanorod arrays by catalyst-free pulsed laser deposition methods.Chem Phys Lett 2004, 396:21-26.Zhang HZ, Sun XC, Wang RM, Yu DP: Growth and formation mechanism of c-oriented ZnO nanorod arrays deposited on glass.JCryst Growth 2004, 269:464-471.
6. Huang MH, Mao S, Feick H, Yan H, Wu Y, Kind H, Weber E, Russo R, Yang P: Room-temperature ultraviolet nanowire nanolasers.Science 2001, 292:1897-1899.
7. Park WI, Kim DH, Jung SW, Yi GC: Metalorganic vapor-phase epitaxial growth of vertically well-aligned ZnO nanorods.Appl Phys Lett 2002, 80:4232-4234.Yan M, Zhang HT, Widjaja EJ, Chang RPH: Self-assembly of well-aligned gallium-doped zinc oxide nanorods.J Appl Phys 2003, 94:5240-5246. Henley SJ, Ashfold MNR, Nicholls DP, Wheatley P, Cherns D: Controlling the size and alignment of ZnO microrods using ZnO thin film templates deposited by pulsed laser ablation. Appl Phys Mater Sci Process 2004, 79:1169-1173.
8. Peterson RB, Fields CL, Gregg BA: Epitaxial chemical deposition of ZnO nanocolumns from NaOH solutions. Langmuir 2004, 20:5114-5118. |Hung CH, Whang WT: Low-temperature solution approach toward highly aligned ZnO nanotip arrays.J Cryst Growth 2004, 268:242-248.Wang L, Zhang X, Zhao S, Zhou G, Zhou Y, Qi J: Synthesis of well-aligned ZnO nanowires by simple physical vapor deposition on c-oriented ZnO thin films without catalysts or additives. Appl Phys Lett 2005, 86:024108-024110.Li Q, Kumar V, Li Y, Zhang H, Marks TJ, Chang RPH: Fabrication of ZnO nanorods and nanotubes in aqueous solutions.Chem Mater 2005, (17):1001-1006.
9. Meulenkamp EA: Synthesis and growth of ZnO nanoparticles.J Phys Chem B 1998, (102):5566-5572.

10. Sakohara S, Ishida M, Anderson MA: Visible luminescence and surface properties of nanosized ZnO colloids prepared by hydrolyzing zinc acetate.J Phys Chem B 1998, (102):10169-10175.
11. McBride RA, Kelly J, McCormack DE: Growth of well-defined ZnO microparticles by hydroxide ion hydrolysis of zinc salts.J Mater Chem 2003, 13:1196-1201.Greene LE, Law M, Tan DH, Montano M, Goldberger J, Somorjai G, Yang P: General route to vertical ZnO nanowire arrays using textured ZnO seeds.Nano Letters 2005, 5:1231-1236. |Weintraub B, Deng Y, Wang ZL: Position-controlled seedless growth of ZnO nanorod arrays on a polymer substrate via wet chemical synthesis.J Phys Chem C 2007, 111:10162-10165.Zhang J, Sun L, Yin J, Su H, Liao C, Yan C: Control of ZnO morphology via a simple solution route.Chem Mater 2002, 14:4172-4177.

Chapter 3

Ammonium Fluoride-activated Synthesis of Cubic δ-TaN Nanoparticles at Low Temperatures

Young-Jun Lee[1], Dae-Young Kim[1], Kap-Ho Lee[2], Moon-Hee Han[1], Kyoung-Soo Kang[3], Ki-Kwang Bae[3], and Jong-Hyeon Lee[1,2]

[1]Graduate School of Green Energy Technology, Chungnam National University, Daejeon 305-764, Republic of Korea

[2]Graduate School of Department of Metallurgical Engineering, Chungnam National University, Daejeon 305-764, Republic of Korea

[3]Korea Institute of Energy Research, 152 Gajeong-ro, Yuseong-gu, Daejeon 305-343, Republic of Korea

ABSTRACT

Cubic delta-tantalum nitride (δ-TaN) nanoparticles were selectively prepared using a $K_2TaF_7 + (5 + k) NaN_3 + kNH_4F$ reactive mixture

(k being the number of moles of NH_4F) via a combustion process under a nitrogen pressure of 2.0 MPa. The combustion temperature, when plotted as a function of the number of moles of NH_4F used, was in the range of 850°C to 1,170°C. X-ray diffraction patterns revealed the formation of cubic δ-TaN nanoparticles at 850°C to 950°C when NH_4F is used in an amount of 2.0 mol (or greater) in the combustion experiment. Phase pure cubic δ-TaN synthesized at k = 4 exhibited a specific surface area of 30.59 m^2/g and grain size of 5 to 10 nm, as estimated from the transmission electron microscopy micrograph. The role of NH_4F in the formation process of δ-TaN is discussed with regard to a hypothetical reaction mechanism.

BACKGROUND

Among the various transition-metal nitrides, TaN is a material that has potential for application in microelectronic components such as capacitors, thin-film resistors, and barrier materials that prevent the diffusion of copper into silicon [1,2]. In addition, TaN has been used in high-temperature ceramic pressure sensors because of its good piezoresistive properties [3]. Also, it is an attractive histocompatible material that can be used in artificial heart valves [4]. Among the various tantalum nitride phases, cubic delta-tantalum nitride (δ-TaN), with a NaCl-type structure (space group: Fm3m), exhibits excellent properties such as high hardness, stability at high temperature, and superconductivity [5].

In general, it is difficult to produce δ-TaN under ambient conditions since its formation requires high temperature and nitrogen pressure. According to the data reported in another study [6], δ-TaN is normally made at more than 1,600°C and 16 MPa of nitrogen pressure. Kieffer et al. synthesized cubic TaN by heating hexagonal TaN above 1,700°C at a N_2 pressure of 6 atm [7]. Matsumoto and Konuma were successful in producing cubic TaN by heating hexagonal TaN at a reduced pressure using a plasma jet [8]. Mashimo et al. were able to transform hexagonal TaN into cubic TaN by both static compression and shock compression at high temperature [9]. Cubic TaN in powder form was also synthesized by self-propagating high-temperature synthesis technique [10,11]. In this process, the combustion of metallic tantalum from 350 to 400 MPa of nitrogen pressure resulted in micrometer size δ-TaN at a temperature above 2,000°C.

More recently, two approaches, solid-state metathesis reaction and nitridation-thermal decomposition [12-14], were adopted for the synthesis of nanosized particles of δ-TaN. O'Loughlin et al. used the metathesis reaction of $TaCl_5$ with Li_3N and 12 mol of NaN_3 to produce δ-TaN [12]. The authors concluded that significant nitrogen pressure created by the addition of NaN_3 enabled cubic-phase TaN to form, along with hexagonal Ta_2N. Solid-state metathesis reaction applied to the $TaCl_5$-Na-NH_4Cl mixture resulted in a bi-phase product at 650°C comprising both hexagonal and cubic phases of TaN [13]. More recently, Liu et al. reported the synthesis of cubic δ-TaN through homogenous reduction of $TaCl_5$ with sodium in liquid ammonia, with a subsequent annealing process at 1,200°C to 1,400°C under high vacuum [14]. Nitridation-thermal decomposition, a two-step process for the synthesis of cubic δ-TaN, was also reported [15]. In the first step, nanosized Ta_2O_5 was nitrided at 800°C for 8 h under an ammonia flow. The as-prepared product was then thermally decomposed at 1,000°C in nitrogen atmosphere, and cubic nanocrystalline δ-TaN was obtained.

In most cases, the products prepared by the above-mentioned methods were often mixtures containing other compounds such as $TaN_{0.5}$ or other nonstoichiometric phases. Therefore, synthesis inefficiency of cubic δ-TaN nanoparticles by known approaches coupled with the multiphase composition of products makes this topic challenging and scientifically attractive.

In this paper, an attractive and rapid approach for synthesizing cubic δ-TaN nanoparticles is developed. This approach includes the combustion of $K_2TaF_7 + (5 + k) NaN_3 + kNH_4F$ exothermic mixture under nitrogen atmosphere and water purification of final products to produce cubic δ-TaN. The approach described in this study is simple and cost-effective for the large-scale production of δ-TaN.

METHODS

For sample preparation, the following chemicals were used: K_2TaF_7 (prepared at the Graduate School of Green Energy Technology, Chungnam National University, Korea), NaN_3 powder (99.0% purity; particle size < 50 μm; Daejung Chemical and Metals Co., Ltd., Shiheung City, South Korea). Chemical-grade ammonium halides (NH_4F and NH_4Cl) were purchased from Samchun Pure Chemical Co., Ltd., Pyeongtaek City, South Korea. All salts were handled in a glove box in dry argon atmosphere (99.99%; Messer, Northumberland, UK).

To prepare the reaction mixture, controlled amounts of reactant powders, K_2TaF_7, NaN_3, and NH_4F, were weighed and thoroughly mixed in a glove box in argon atmosphere. About 60 to 80 g of the mixture was compacted by hand in a stainless steel cup (4.0 cm in diameter) and placed in a high-pressure reaction vessel for combustion (Figure 1). A vacuum was applied to remove the air from the combustion vessel, which was then filled with nitrogen gas with a pressure of 2.0 MPa. The combustion process was initiated by a hot nickel-chromium filament system, and the reaction temperatures were measured using WR-20/WR-5 thermocouples inserted into the reaction pellet. After completion of the combustion process, the burned-down sample was cooled to room temperature and transferred to a 500-ml beaker for further purification. The sample was purified by washing with distilled water in order to remove the NaF and KF salts that formed during the reaction. The purified black powder was dried in air at 80°C to 90°C.

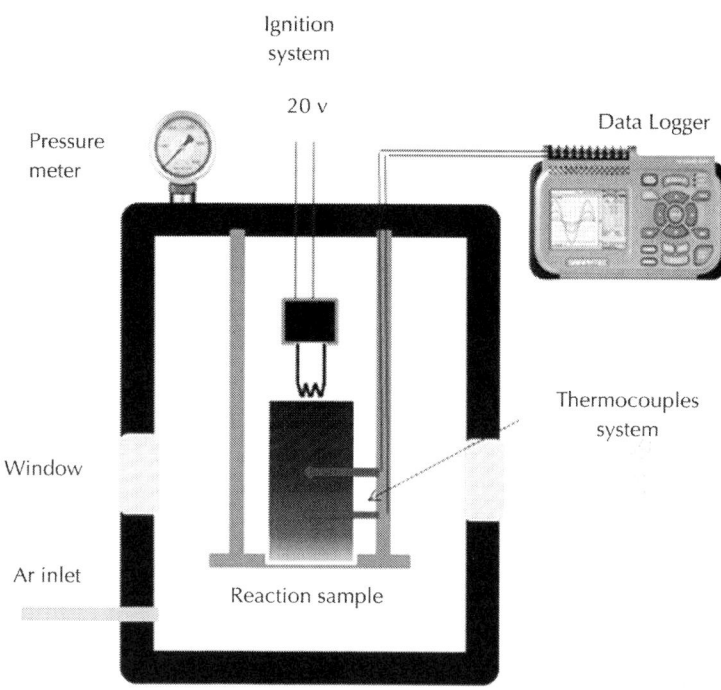

Figure 1: Experimental setup for the synthesis cubic TaN nanoparticles.

We used the simulation software 'Thermo' to predict adiabatic combustion temperature (T_{ad}) and concentrations of equilibrium phases in the combustion wave [16]. Calculations of equilibrium characteristics were based on minimizing the thermodynamic potential of the system. The initial parameters (temperature and pressure) of the system were set as 25°C and 2.0 MPa, respectively.

The crystal structure and morphology of the TaN nanoparticles were characterized X-ray diffraction (XRD) with Cu Kα radiation (D5000, Siemens AG, Munich, Germany), field-emission scanning electron microscopy (FESEM; JSM 6330F, JEOL Ltd., Akishima, Tokyo, Japan), and transmission electron microscopy (TEM; JEM 2010, JEOL Ltd.). The specific surface area of the nanoparticles was determined from the linear portion of the Brunauer, Emmett, and Teller plot.

RESULTS AND DISCUSSION

Results

Adiabatic Combustion Temperature and Equilibrium Phases

The combustion thermodynamics of the $K_2TaF_7 + (5 + k) NaN_3 + kNH_4F$ system calculated by the Thermo simulation software is shown Figure 2. This calculation provides equilibrium product concentration (C) and T_{ad} as a function of the number of moles of NH_4F used (k). As shown in Figure 2, the calculated adiabatic combustion temperature shows an almost linear decreasing tendency with increasing k. The mixture with the highest temperature, near 1,425°C, is predicted for $K_2TaF_7 + 5NaN_3$ binary mixture ($k = 0$). As estimated from Figure 2, the temperature change from 1,425°C to 1,000°C is observed when k changes from 0 to 5. The reaction products predicted by thermodynamic analysis comprise solid tantalum nitride (TaN), liquid fluorides of alkaline metal (NaF, KF), and gaseous H_2 and N_2. The concentration of TaN and KF predicted by thermodynamic analysis is constant in the given interval of NH_4F, whereas the concentration of NaF, H_2, and N_2 has been increasing with increasing k. Intensive gas release in the designed system, especially at higher k, may generate high pressure in the combustion wave. Our estimation shows that the pressure in the combustion wave may reach tens and even hundreds of atmospheres. This can be very helpful to accelerate the formation of cubic phase TaN at given temperatures. This also indicates that one must keep external nitrogen pressure relatively high to prevent distortion of the sample during the combustion experiment and to avoid the scattering of reaction mass inside of the combustion chamber. Therefore, the data obtained from thermodynamic analysis can serve as a good theoretical guideline for controlling the combustion process and optimizing the synthesis conditions of cubic TaN nanoparticles.

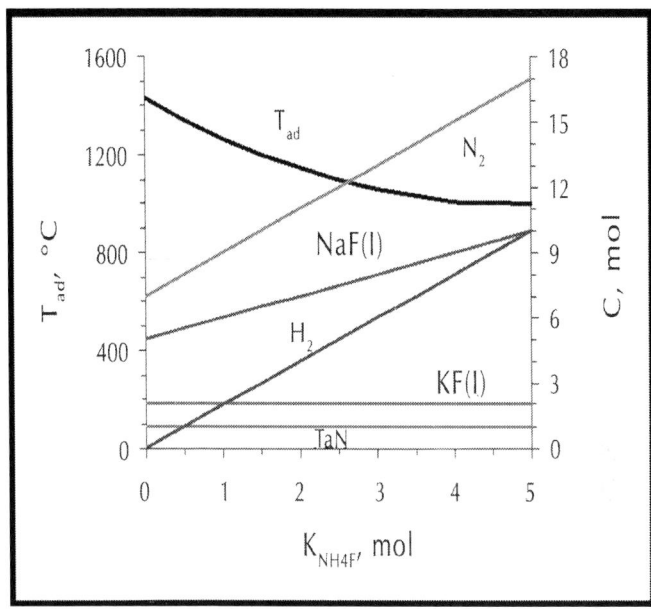

Figure 2: T_{ad} and equilibrium phases in $K_2TaF_7 + (5 + k) NaN_3 + kNH_4F$ system upon k.

DSC-TGA Curves and Combustion Parameters

Differential scanning calorimetry (DSC) and thermogravimetric analysis (TGA) were carried out in order to elucidate the thermal behavior of the $K_2TaF_7 + 5NaN_3$ (C1) and $K_2TaF_7 + 5NaN_3 + 4NH_4F$ (C$_2$) reaction mixtures as well as to determine the weight losses incurred during the heating process. The samples were heated at a rate of 20°C/min in a flow of argon gas. The weight loss for both samples is in the range from approximately 60°C to 380°C (Figure 3, lines 1 and 1') which is mainly caused by the decomposition of NH_4F and NaN_3. Therefore, above 380°C, no drop of mass was recorded by TGA analysis. The highest maximum of DSC signals (Figure 3, lines 2 and 2') is reached at 330°C and 380°C. This means that at the given temperatures, a strong exothermic reduction of K_2TaF_7 by Na has occurred in the C1 and C2 mixtures, resulting in large outflow of heat and sharp weight losses. In addition, the exothermic peak at round 330°C (mixture C1) is significantly higher than the exothermic peak recorded at around 380°C for C2. This indicates that the NH_4F-containing mixture is less exothermic, and this finding is concise with the thermodynamic analysis data. Therefore, it can be suggested that in the given systems, the combustion process in the $K_2TaF_7 + (5 + k)NaN_3 + kNH_4F$ system starts at around 350 ± 50°C.

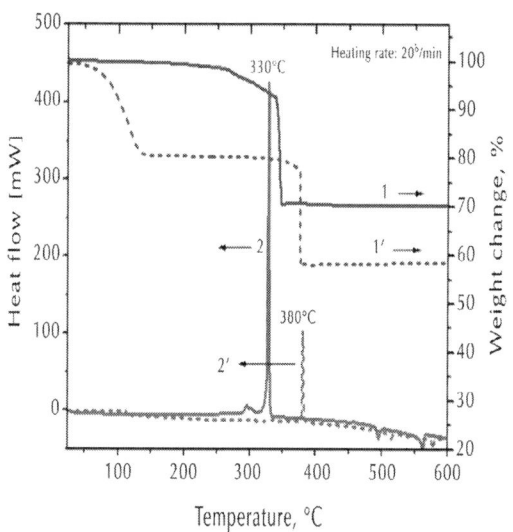

Figure 3: DSC-TGA curves of $K_2TaF_7 + 5NaN_3$ and $K_2TaF_7 + 9NaN_3 + 4NH_4F$ systems in argon atmosphere.

Figure 4 shows the temperature-time profiles for the combustion wave of the $K_2TaF_7 + (5 +k)NaN_3 + kNH_4F$ mixture over the reaction time (t). As shown in Figure 4, the starting temperature for the combustion process is denoted by T^* (350 ± 50°C) and corresponds to the sharp peaks in the DSC curve (Figure 3). One can see that in the beginning of the reaction zone, the temperature increases rapidly from 25°C to 700°C and then to 1,000°C, and then long-tailed post-combustion processes followed. The combustion temperature (T_c) showed a tendency to decrease with the amount of NH_4F used. In the investigated interval of k, the T_c drops from 1,170°C ($k = 0$) to 850°C ($k = 4$). The maximum combustion velocity ($U_c = 0.5$ cm/s) occurred at the nearly stoichiometric mixture ($k = 0$), but combustion velocity decreased significantly as the $K_2TaF_7 + 5NaN_3$ mixture became 'diluted' with NH_4F.

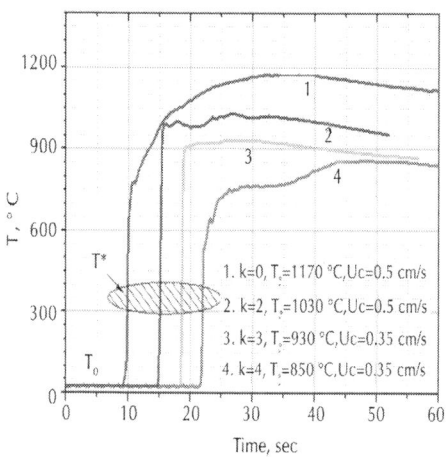

Figure 4: Temperature-time profiles in $K_2TaF_7 + (5 + k)NaN_3 + kNH_4F$ system.

Characteristics of Combusted Samples and Powders

Figure 5 shows photographs of the as-combusted (Figure 5a, b) and water-purified (Figure 5c) samples. After combustion, the sample of the $K_2TaF_7 + 5NaN_3$ composition ($k = 0$) retained its original shape and size (Figure 5a). However, the samples produced using 2.0 to 4.0 mol of NH_4F had melted after the combustion process, forming a brown-colored, brittle, and shapeless molten product. For instance, several fragments of the sample prepared with $k = 4$ are shown in Figure 5b. Many large pores, due to the release of N_2 and H_2 gases during the combustion process, can be seen in the solid molten mass. After dissolving alkali fluorides (NaF and

KF) into warm distillated water, TaN fine powders were obtained. A photograph of finally purified TaN samples prepared from the $K_2TaF_7 + 5NaN_3 + 4NH_4F$ mixture is shown in Figure 5c. Its color is uniformly black, and specific gravity lies between 0.7 and 0.9 g/cm^3.

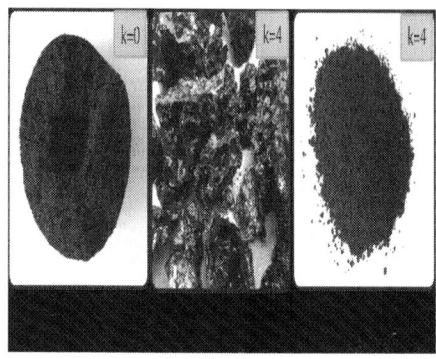

Figure 5: Photographs of as-combusted (a, b) and water-purified (c) samples.

The XRD patterns for the water-purified powders that had been prepared with different amounts of NH$_4$F are shown in Figure 6. Diffraction peaks of the sample prepared at $k = 0$ (without NH$_4$F) indicate three nitride phases: hexagonal ε-TaN, TaN$_{0.8}$, and Ta$_2$N (Figure 6a). The cubic δ-TaN phase was detected in large amounts, along with the ε-TaN and TaN$_{0.8}$ phases for samples with $k = 2$ (Figure 6b). By applying 4 mol of NH$_4$F to the reaction of K$_2$TaF$_7$ and NaN$_3$, the only crystalline product produced is cubic TaN. The diffraction peaks marked in Figure 6c correspond to face-centered cubic TaN (JCPDS 32–1283). Thus, the reaction of K$_2$TaF$_7$, NaN$_3$, and NH$_4$ (in a 1:9:4 molar ratio) results in the formation of phase-pure crystalline cubic TaN. The peaks for δ-TaN are weak and broad, indicating the small size of its particles. The lattice parameter calculated from the highest intensity peak (111) was $a = 4.32$ Å. This was in good agreement with the previously reported value of 0.433 ± 0.001 nm for thin films [17]. The nitrogen content in the powders at various k values is shown in Table 1. It shows that the nitrogen content at $k = 0$ is 7.01%, which corresponds to the $TaN_{0.98}$ composition. With increasing k, the nitrogen content then slowly drops down, reaching to 6.51% at $k = 4$. This amount of nitrogen theoretically corresponds to the $TaN_{0.91}$ composition. All the powders contain about 0.15% carbon.

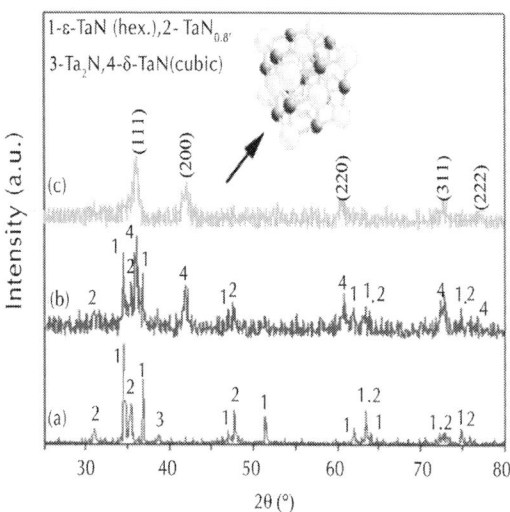

Figure 6: XRD patterns of water-purified powders synthesized from K_2TaF_7 + $(5 + k)NaN_3$ + kNH_4F mixture. (a) $k = 0$, (b) $k = 2.0$, and (c) $k = 4.0$.

Table 1: Content of nitrogen in TaN

k (mol)	N (%)	Formula
0	7.01	$TaN_{0.98}$
2	6.95	$TaN_{0.97}$
3	6.72	$TaN_{0.94}$
4	6.51	$TaN_{0.91}$

Lee et al.

Lee et al. *Nanoscale Research Letters* 2013 **8**:126, doi:10.1186/1556-276X-8-126

The SEM microstructure of the combustion product ($k = 0$) right after the synthesis process is shown in Figure 7a. Due to a large portion of molten fluorides (5NaF to 2KF), the final product has a molten microstructure in which the crystalline particles of tantalum nitride are embedded. The microstructure of the same sample after water purification is shown in Figure 7b. A part of TaN particles were crystallized in a rodlike fashion; at that, the length of rods is about 0.5 to 1.5 μm, as estimated from the micrograph. A large portion of small particles whose sizes are on the order of submicrometers also exist on the same micrograph. We think that the presence of different-sized particles in Figure 7b can be associated with the phase composition of the product, i.e., the rod-shaped particles most likely are those of hexagonal ε-TaN, whereas the small-sized particles belong to the $TaN_{0.8}$ and Ta_2N phases. With an increase in k, the rod-shaped particles disappeared, and the size of particles became smaller and uniform.

Figure 7: SEM micrographs of reaction product (a), and water-purified TaN samples with EDX analysis (b, c). (a) $k = 0$, (b) $k = 0$, and (c) $k = 4$.

As a typical example, the micrograph of the cubic δ-TaN particles produced using 4.0 mol of NH_4F is shown in Figure 7c. These particles are less than 100 nm in size. They usually exist in the form of relatively large clusters (0.5 to 1.0 μm), owing to the attractive forces between the particles. EDS analysis taken from rodlike and small-sized particles (Figure 7b,c) shows Ta and N as the main elements; however, small peaks of oxygen also exist.

Figure 8a shows the TEM image and the corresponding selected area electron diffraction (SAED) pattern of the cubic δ-TaN nanoparticles synthesized at 800°C from the K_2TaF_7 + 9NaN_3 + 4NH_4F mixture. The TEM image confirmed the formation of TaN nanoparticles, which had an average size of <10 nm. From the diffraction rings in the SAED pattern, shown in the inset of Figure 8a, (111), (200), and (220) planes were identified in the δ-TaN nanoparticles. The average spacing between the stacks was 2.5 to 2.6 Å (111), as estimated from the HRTEM image (Figure 7b).

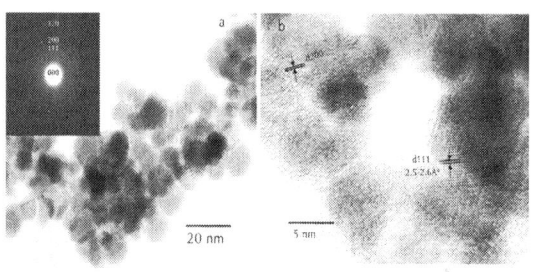

Figure 8: TEM micrograph (a), SAED pattern (inset of a), and HRTEM image (b) of cubic TaN nanoparticles.

DISCUSSION

The phase-pure cubic TaN nanoparticles reported here have proven to be difficult to synthesize in previous attempts using solid-state metathesis reactions [12-14]. However, our experimental results clearly indicate that cubic-phase δ-TaN nanoparticles can be produced at moderate temperatures, within several or tens of seconds by combustion of the K_2TaF_7 + (5 + k) NaN_3 +kNH_4F mixture under 2.0 MPa of nitrogen pressure. The entire combustion process, with the optimized NH_4F amount used (4.0 mol), can be presented as follows:

$$K_2TaF_7 + 9NaN_3 + 4NH_4F \rightarrow 2KF + 9NaF + TaN + 15N_2 + 8H_2 \quad (1)$$

As shown above, the forming of cubic TaN from the exothermic mixture of K_2TaF_7 + 5NaN_3 composition does not occur despite a relatively high combustion temperature (1,170°C). Under conditions, however, the addition of ammonium fluoride to the reaction mixture had a favorable effect on the cubic-phase δ-TaN nanoparticle synthesis, despite large drops in the combustion temperature (850°C; k = 4). The replacement of NH_4F with NH_4Cl slightly lowered the combustion temperature to 850°C (k = 4). However, cubic-phase δ-TaN nanoparticles were obtained. Therefore, the addition of ammonium halides to the combustion reaction can provide low pressure and temperature route for the synthesis of the cubic TaN.

Ammonium halides appear to have two functions: acting first as a heat sink and then as a source of nitrogen and hydrogen. According to Equation 1, each mole of NH_4F added to the mixture required 1.0 mol of NaN_3 in order to neutralize HF, which forms after the decomposition of NH_4F. Therefore, the intermediate gas phase products of the combustion process may consist of NH_3, N_2, and H_2. However, at higher combustion temperatures (>500°C), a decomposition of NH_3 occurs, and N_2 and H_2 gases become dominant. A simple estimation from Equation 1 shows that the total amounts of N_2 and H_2 in the combustion wave are 15.5 and 8 mol, respectively. We think that the presence of N_2 and H_2 gases in the combustion wave is the key factor, making cubic TaN formation favorable. In order to prove this assumption, we have prepared a hydrogen-free mixture of K_2TaF_7 + 5.175ZnF_2 + 10.35 NaN_3 composition and combusted under 2.0 MPa nitrogen pressure. The combustion process in the given system can be presented as follows:

$$K_2TaF_7 + 5.175ZnF_2 + 10.35NaN_3 \rightarrow 2KF + 10.35NaF + TaN + 15N_2 + 5.175Zn \quad (2)$$

In this process, the total amount of NaN_3 was set at 10.35 mol to produce 15.5 mol of N_2, as seen in the reaction (Equation 2). The combustion temperature of the K_2TaF_7 + 5.175ZnF_2 + 10.35

NaN$_3$ mixture measured by thermocouples was 900°C. The reaction product after acid leaching was a black powder and was a component from hexagonal ε-TaN and Ta$_2$N according to XRD analysis. No peaks matching to cubic TaN was found on the XRD patterns. The response of cubic TaN to ammonium halides raised the question about the mechanism of the process. At present, we do not have a clear explanation of the role that ammonium halide has during the synthesis process. However, a plausible hypothesis can be offered with respect to the underlying mechanism. We believe that the hydrogen that is released from ammonium halide may stimulate a process of hydration-dehydration of Ta in the intermediate stages of the combustion process and may lead to vacancies in the tantalum lattice without affecting its crystal structure. These free vacancies created by hydrogen atoms could be easily occupied by nitrogen atoms at higher combustion temperatures, thus leading to the formation of cubic δ-TaN. Another possible explanation for the cubic phase may involve the formation of tantalum amido- or imido-fluorides (Ta(NH$_2$)$_2$F$_3$·4NH$_3$ or Ta(NH$_2$)$_2$F$_4$·6NH$_3$) in a manner similar to the previously reported formation of tantalum amido- or imido-chlorides (Ta(NH$_2$)$_2$Cl$_3$·4NH$_3$ or Ta(NH$_2$)$_2$Cl$_4$·6NH$_3$)[18,19]. However a further, detailed investigation is needed to clarify the mechanism behind the formation of cubic tantalum nitride using ammonium halides.

CONCLUSIONS

Nanocrystalline cubic δ-TaN was prepared by a solid combustion synthesis method using the K$_2$TaF$_7$ + (5 + k)NaN$_3$ + kNH$_4$F reactive mixture. It was shown that without NH$_4$F, the maximum temperature of K$_2$TaF$_7$ + 5NaN$_3$ mixture is 1,170°C, and the combustion product is multiphase consisting of hexagonal TaN as well as TaN$_{0.8}$ and Ta$_2$N phases. However, the addition of NH$_4$F to the reactive mixture stimulates the formation of cubic δ-TaN. Phase-pure cubic δ-TaN was obtained when NH$_4$F in the amount of 4.0 mol (or greater) was used in the combustion experiments. The formation temperatures for cubic δ-TaN were as low as 850°C to 950°C. Cubic δ-TaN synthesized using 4.0 mol of NH_4F exhibited a specific surface area of 30.59 m^2/g and a grain size of 5 to 10 nm, as estimated from its TEM micrograph. The approach developed in this study is a simple and cost-efficient method for the large-scale production of δ-TaN.

AUTHORS' CONTRIBUTIONS

MHH, KHL, KSK, KKB, and JHL conceived the review. YJL performed the experiments with the help from DYK. YJL drafted the manuscript. All authors read and approved the final manuscript.

ACKNOWLEDGMENTS

This research was supported by KIER R&D program (Project number KIER B2-2144-03) under Korea Institute of Energy, Republic of Korea.

REFERENCES

1. Lovejoy ML, Patrizi GA, Rieger DJ, Barbour JC: Thin-film tantalum-nitride resistor technology for phosphide-based optoelectronics. *Thin Solid Films* 1996, 290–291(2):513-517.

2. Laurila , Zeng K, Kivilahti JK, Molarius J, Riekkinen T, Suni I: Tantalum carbide and nitride diffusion barriers for Cu metallization. *Microelectron Eng* 2002, 60(1):71-80.

3. Ayerdi I, Castano E, Garcia-Alonso A, Gracia J: High-temperature ceramic pressure sensor. *Sensors Actuators A* 1997, 60(1):72-75.

4. Leng YX, Sun H, Yang P, Chen JY, Wang J, Wan GJ, Huang N, Tian XB, Wang LP, Chu PK:Biomedical properties of tantalum nitride films synthesized by reactive magnetron sputtering. *Thin Solid Films* 2001, 398–399(2):471-475.

5. Mashimo T, Nishida M, Yamaya S, Yamasaki H: Stoichiometric B1-type tantalum nitride and a sintered body thereof and method of synthesizing, the B1-type of tantalum nitride. *US Patent* April 1994, 5306320:26.
6. Gatterer J, Dufek G, Etmayer P, Kieffer R: The cubic tantalum mononitride (B 1) and its miscibility with the isotypic mononitrides and monocarbides of the 4a and 5a group metals. *Monatch Chem* 1975, 106:1137.
7. Kieffer R, Ettmayer P, Freundhofmeier M, Gatter J: The cubic tantalum mononitride with B1 structure. *Monatsh Chem* 1971, 102:483.
8. Matsumoto O, Konuma M, Kanzaki Y: Formation of cubic tantalum nitride by heating hexagonal tantalum nitride in a nitrogen-argon plasma jet. *J Less Common Met* 1978, 60:147.
9. Mashimo T, Tashiro S, Nishida M, Miyahara K, Eto E: B1-type and WC-type phase bulk bodies of tantalum nitride prepared by shock and static compressions. *Phys B* 1997, 239:13.
10. Petrunin VF, Sorokin NI, Borovinskaya IP, Pityulin AN: Stability of cubic tantalum nitrides during heat treatment. *Powder Metall Met Ceram* 1980, 19:62-64.
11. Merzhanov AG, Borovinskaya IP, Volodin YE: Mechanism of combustion for porous metal specimens in nitrogen. *DANKAS* 1972, 206:905-908.
12. O'Loughlin JL, Wallace CH, Knox MS, Kaner RB: Rapid solid-state synthesis of Ta, Cr, and Mo nitrides. *Inorg Chem* 2001, 40:2240-2245.
13. Shi L, Yang ZH, Chen LY, Qian YT: Synthesis and characterization of nanocrystalline TaN. *Solid State Commun* 2005, 133(2):117-120.
14. Liu L, Huang K, Hou J, Zhu H: Structure refinement for tantalum nitrides nanocrystals with various morphologies. *Mater Res Bull* 2012, 47:1630-1635.
15. Fu B, Gao L: Synthesis of nanocrystalline cubic tantalum(III) nitride powders by nitridation–thermal decomposition. *J Am Ceram Soc* 2005, 88:3519-3521.

16. Shiryaev AA: Thermodynamics of SHS processes: advanced approach. *Int J SHS* 1995, 4:351.
17. Matenoglou GM, Koutsokeras LE, Lekka CE, Abadias G, Camelio S, Evangelakis GA, Kosmidis C, Patsalas P: Optical properties, structural parameters, and bonding of highly textured rocksalt tantalum nitride films. *J Appl Phys* 2008, 104:124907.
18. Holl MB, Kersting M, Pendley BD, Wolczanski PT: Ammonolysis of tantalum alkyls: formation of cubic tantalum nitride and a trimeric nitride, [Cp*MeTaN]3 tris[(.eta.5-pentamethylcyclopentadienyl)(methyl)nitridotantalum]. *Inorg Chem* 1990, 29(8):1518-1526.
19. Choi D, Kumta PN: Synthesis, structure, and electrochemical characterization of nanocrystalline tantalum and tungsten nitrides. *J Am Ceram Soc* 2007, 90(10):3113-3120.

Chapter 4

Wettability Switching Techniques on Superhydrophobic Surfaces

Nicolas Verplanck[1], Yannick Coffinier[1,2], Vincent Thomy[1], and Rabah Boukherroub[1,2]

[1]Institut d'Electronique, de Microélectronique et de Nanotechnologie (IEMN), UMR 8520, Cité Scientifique, Avenue Poincaré, B.P. 60069, 59652, Villeneuve d'Ascq, France

[2]Institut de Recherche Interdisciplinaire (IRI), FRE 2963, Cité Scientifique, Avenue Poincaré, B.P. 60069, 59652, Villeneuve d'Ascq, France

ABSTRACT

The wetting properties of superhydrophobic surfaces have generated worldwide research interest. A water drop on these surfaces forms a nearly perfect spherical pearl. Superhydrophobic materials hold considerable promise for potential applications ranging from self-cleaning surfaces, completely water impermeable textiles to low cost energy displacement of liquids in lab-on-chip devices. However, the dynamic modification of the liquid droplets behavior and in particular of their wetting properties on these surfaces is still a challenging issue. In this review, after a brief overview on superhydrophobic states definition, the techniques leading to the modification of wettability behavior on superhydrophobic surfaces under specific conditions: optical, magnetic, mechanical, chemical, thermal are discussed. Finally, a focus on electrowetting is made from historical phenomenon pointed out some decades ago on classical planar hydrophobic surfaces to recent breakthrough obtained on superhydrophobic surfaces.

INTRODUCTION

Biological surfaces, like lotus leaves, exhibit the amazing property for not being wetted by water leading to a self-cleaning effect. The lotus leaves capability to remain clean from dirt and particles is attributed to the superhydrophobic nature of the leaves surface. The latter is composed of micro and nano structures covered with a hydrophobic wax, creating a carpet fakir, where water droplets attained a quasi-spherical shape. In order to mimic these properties, artificial superhydrophobic surfaces have been prepared by several means, including the generation of rough surfaces coated with low surface energy molecules [1-6], roughening the surface of hydrophobic materials [7-9], and creating well-ordered structures using micromachining and etching methods [10,11].

However, the modification of the liquid droplets behavior and in particular of their wetting properties on these surfaces is still a challenging issue. Functional surfaces with controlled wetting properties, which can respond to external stimuli, have attracted huge interest of the scientific community due to their wide range of potential applications, including microfluidic devices, controllable drug delivery and self-cleaning surfaces. In this review, after a brief overview on superhydrophobic states definition, we will discuss the techniques leading to the modification of wettability behavior on superhydrophobic surfaces under specific conditions: optical, magnetic, mechanical, chemical, thermal... Finally, a focus on electrowetting will be made from historical phenomenon pointed out some decades ago on classical planar hydrophobic surfaces to recent breakthrough obtained on superhydrophobic surfaces.

SURFACE WETTING

Introduction

The wetting property of a surface is defined according to the angle θ, which forms a liquid droplet on the three phase contact line (interface of three media—Fig. 1a). A surface is regarded as wetting when the contact angle, which forms a drop with this one, is lower than 90° (Fig. 1a). In the opposite case (the contact angle is higher than 90°), the surface is nonwetting (Fig. 1b). For water, the terms "hydrophilic" and "hydrophobic" are commonly used for wetting and nonwetting surfaces, respectively.

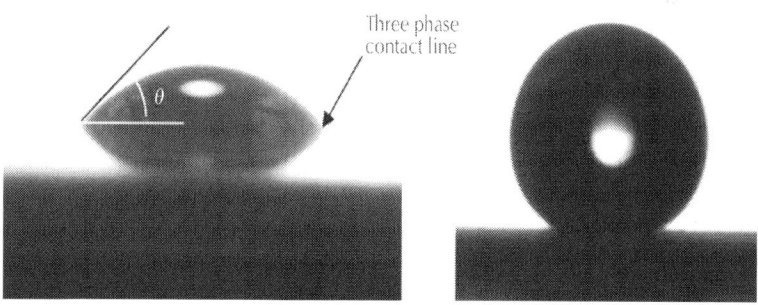

Figure 1: Droplet of water deposited on two surfaces of different energies: (a) wetting surface (θ < 90°), (b) nonwetting surface (θ > 90°).

The contact angle of a liquid on a surface according to the surface tension is given by the relation of Young (1). The surface tension, noted γ, is the tension which exists at the interface of two systems (solid/liquid, liquid/liquid, solid/gas). It is expressed in energy per unit of area (mJ m^{-2}), but can also be regarded as a force per unit of length (mN m^{-1}). From this definition, it is possible to identify three forces acting on the three phase contact line: γ_{LG} (liquid surface stress/gas), γ_{LS} (liquid/solid surface stress) and γ_{SG} (solid surface stress/gas). The three forces are represented in Fig. 2.

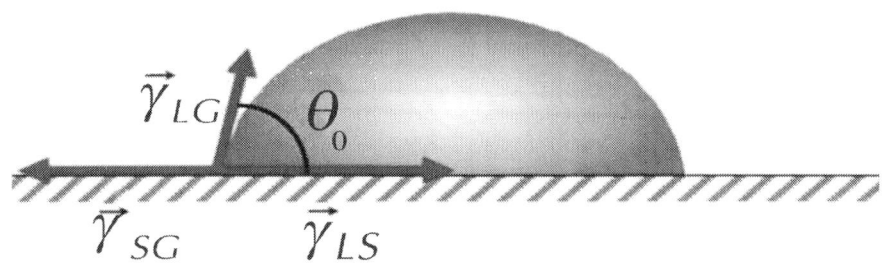

Figure 2: Surface forces acting on the three phase contact line of a liquid droplet deposited on a substrate.

At the equilibrium state:

$$\vec{\gamma}_{LS} + \vec{\gamma} + \vec{\gamma}_{SG} = 0$$

By projection on the solid, the relation of Young [12] is obtained:

$$\gamma_{LS} = \gamma_{SG} - \gamma \cos \theta_0 \qquad (1)$$

It is also possible to establish the Eq. 1 by calculus of the surface energy variation related to a displacement dx of the three phase contact line:

$$dE = (\gamma_{LS} - \gamma_{SG})dx + \gamma dx \cos \theta$$

At the equilibrium state, using energy minimization (dE = 0), the Young relation (1) is found. This approach will be used thereafter to determine the relations of Wenzel and Cassie–Baxter on superhydrophobic surfaces.

Concretely, following the rule of Zisman [13, 14], wetting surfaces are surfaces of high energy (~500–5,000 mN m^{-1}), where the chemical binding energies are about an eV (ionic, covalent, metal connections). The wetting materials are typically oxides (glass), metal oxides… On the other hand, nonwetting surfaces are characterized by low surface energy (~10–50 mN m^{-1}). For these materials, the binding energies are about kT (ex: crystalline substrates and polymers) [15].

Hysteresis

The hysteresis of a surface is related to its imperfections. Indeed, the formula of Young considers that there is only one contact angle, the static contact angle, and noted θ_0. However, this configuration exists only for perfect surfaces. Generally, surfaces present imperfections related to physical defects like roughness or to chemical variations. The static contact angle thus lies between two values called advanced angle, noted θ_A, and receding angle, noted θ_R. The difference between these two angles ($\theta_A - \theta_R$) is called hysteresis. While this force is opposed to droplet motion, the smaller hysteresis is, the more it will be easy to move the liquid droplet. Concretely, these angles can be measured thanks to the shape of a droplet on a tilted surface (Fig. 3).

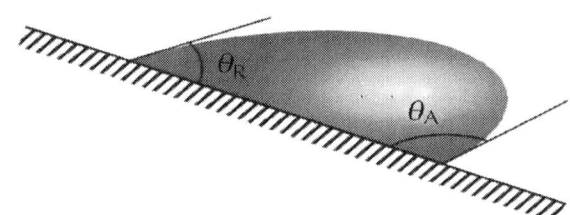

Figure 3: Advanced θA and receding θR angles of a liquid droplet on a tilted surface.

WETTING ON SUPERHYDROPHOBIC SURFACES: WENZEL AND CASSIE–BAXTER STATES

The lotus leaves are known for their water repellency and consequently to remain clean from any parasitic dust or debris. This phenomenon (also called rolling ball state) is very common in nature not only for the lotus, but also for nearly 200 other species: vegetable and animal like species. For example, the wings of a butterfly are covered with shapes whose size and geometrical form lead to a superhydrophobic state and are at the origin of their color (Fig. 4).

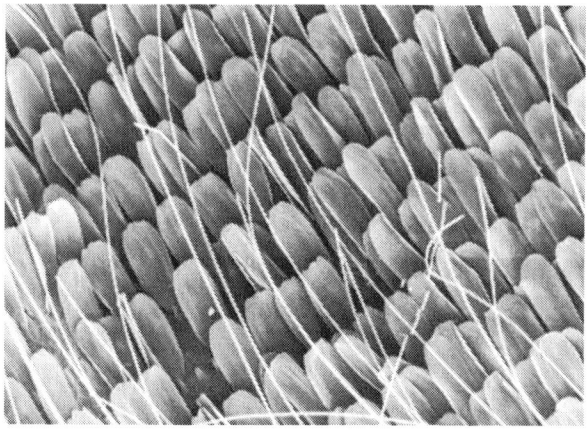

Figure 4: SEM image of a butterfly wings [16]. Reprinted with permission. Copyright of the University of Bath (UK).

The common point between all these surfaces is their roughness. Indeed, the surfaces are composed of nanometric structures limiting the impregnation of the liquid and pushing back the drop. Most of the time, the surfaces are made of a second scale of roughness, consisting of micrometric size. In order to minimize its energy, a liquid droplet forms a liquid pearl on the microstructured surface. The superhydrophobicity term is thus used when the apparent contact angle of a water droplet on a surface reaches values higher than 150°.

Previously, the studied substrates were regarded as smooth surfaces, i.e. the roughness of the substrate was sufficiently low and thus does not influence the wetting properties of the surface. In this case, the relation of Young (1) gives the value of the contact angle on the surface (which we will henceforth call angle of Young). However, a surface can have a physical heterogeneity (roughness) or a chemical composition variation (materials with different surface energies). In this case, a drop deposited on the surface reacts in several ways. A new contact angle is then observed, called apparent contact angle and generally noted *. It should be noticed that locally, the contact angle between the liquid droplet and the surface are always the angle of Young. Two models exist: the model of Wenzel [17,18] and of Cassie–Baxter [19].

These two models were highlighted by the experiment of Johnson and Dettre [20]. Many research teams have tried to understand in more detail the superhydrophobicity phenomenon [21] and particularly the difficulty of the wetting transition from Wenzel to Cassie configuration [22]. A drop on a rough and hydrophobic surface can adopt two configurations: a Wenzel [23] (complete wetting) and a Cassie–Baxter configuration (partial wetting), as presented in Fig. 5a and b, respectively. In both cases, even if locally, the contact angle does not change (angle of Young), an increase in the apparent contact angle * of the drop is observed.

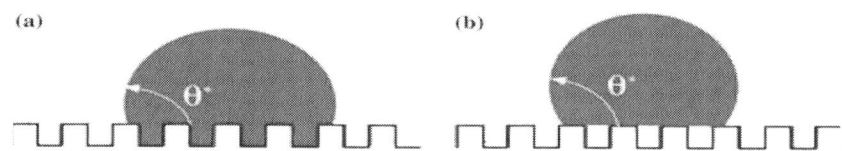

Figure 5: Superhydrophobic surfaces: (a) Wenzel, (b) Cassie–Baxter model [24]. Reprinted with permission from [24]. Copyright 2007 Royal Society of Chemistry.

For a superhydrophobic surface, the fundamental difference between the two models is the hysteresis value. The first experiment on this subject was conducted by Johnson and Dettre (1964) who measured the advancing and receding contact angles, according to the surface roughness [20]. For a low roughness, a strong hysteresis being able to reach 100° (Wenzel) is observed and attributed to an increase in the substrate surface in contact with the drop. Starting from a certain roughness (not quantified in their experiment), the hysteresis becomes quasi null resulting from the formation of air pockets under the drop. The receding angle approaches the advancing angle.

Other experiments also show that for a drop, in a Cassie–Baxter state, it is possible to obtain a contact angle quite higher than for a drop in Wenzel state (Fig. 6a) [24]. The drop on the left is in a Cassie–Baxter state whereas the drop on the right is in a Wenzel state. After partial evaporation of the drop (Fig. 6b), the observed angle (which is the receding angle) is similar to the advancing angle for the drop on the left whereas the drop on the right appears like trapped on a hydrophilic surface.

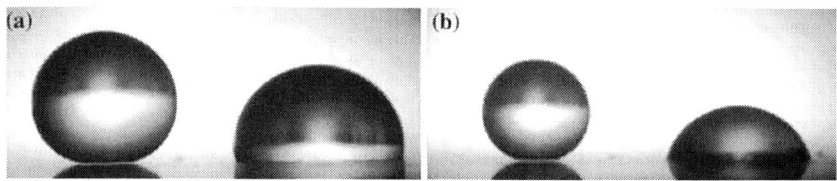

Figure 6: Illustration of the difference between the Cassie–Baxter and Wenzel states: (a) after deposition of the liquid drops on the surface, (b) after evaporation [24]. Reprinted with permission from [24]. Copyright 2007 Royal Society of Chemistry.

In the following two paragraphs, we will discuss in detail the two models. Then we will show that the reality is more complex, in particular in the presence of metastable states in the Cassie–Baxter model.

Wenzel (1936)

When a surface exhibits a low roughness, the drop follows the surface and is impaled on roughness (Fig. 5a). In this case, the solid surface/liquid and solid/gas energies are respectively $r\gamma_{SL}$ and $r\gamma_{SG}$, where the roughness r is defined as the relationship between real surface and apparent surface (r > 1 for a rough surface, and r = 1 for a perfectly smooth surface) [25]. A dx displacement of the three phase contact line thus involves a variation of energy:

$$dE = r(\gamma_{SL} - \gamma_{SV})dx + \gamma dx \cos\theta^* \quad (2)$$

At the equilibrium state (dE = 0), for a null roughness, i.e. for r = 1, we find the relation of Young. For a nonnull roughness, the relation of Wenzel [18] is obtained:

$$\cos\theta^* = r\cos\theta \quad (3)$$

The question is to know what the conditions to be in this configuration are. In this relation, the angle of Young cannot be modulated since on a planar surface the optimal contact angle value is around 120° for water. Moreover, this relation implies that it is possible to reach an apparent contact angle of 180° as soon as the product rcos reaches −1 (as shown in Fig. 7). However an apparent angle * of 180° cannot be observed because the drop must preserve a surface of contact with the substrate. Thus the only parameter that can be modulated is the roughness. However, a strong roughness involves a configuration of Cassie–Baxter. Indeed, a liquid droplet rather minimizes its energy while remaining on a surface of a strong roughness than penetrating in the asperities. So the law of Wenzel is valid only for one certain scale of roughness and thus for apparent angles lower than 180°.

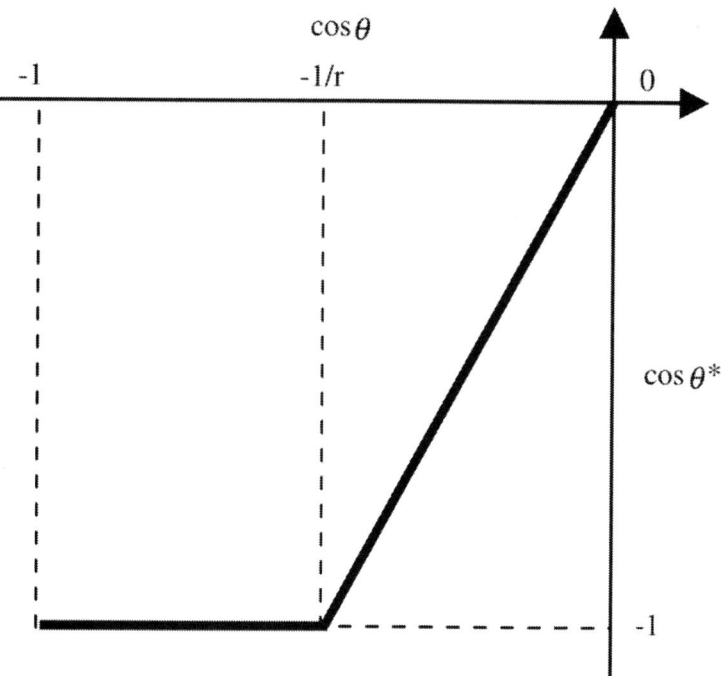

Figure 7: Apparent contact angle according to the angle of Young (relation of Wenzel).

In this type of behavior, the liquid/solid interface and the hysteresis are strongly increased. The drop sticks to the surface and the Wenzel state contrasts with the superhydrophobicity idea i.e. the rolling ball effect.

Cassie–Baxter (1944)

Cassie and Baxter did not directly investigate the wetting behavior of liquid droplets on superhydrophobic surfaces. They were more particularly interested in planar surfaces with chemical heterogeneity (Fig. 8).

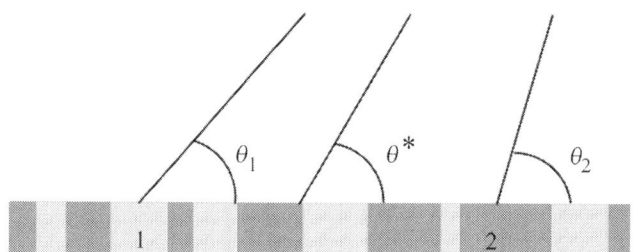

Figure 8: Planar surface composed of two different and chemically heterogeneous materials.

The examined surface consists of two materials; each one has its own surface energy, characteristic contact angle, and occupies a definite fraction of the surface. If material 1 is hydrophobic and material 2 is replaced by air, a drop in contact with each of the two phases (solid and air) forms respective contact angles θ_E and 180°, whereas the fractions of respective surfaces are Φ_S and $(1 - \Phi_S)$. Considering a displacement dx of the three phase contact line, the change of energy dE could be expressed by:

$$dE = \phi_S(\gamma_{SL} - \gamma_{SV})dx + (1 - \phi_S)\gamma dx + \gamma dx \cos\theta^* \tag{4}$$

By using the relation of Young, the minimum of E leads to the Cassie–Baxter relation:

$$\cos\theta^* = -1 + \phi_S(\cos\theta_E + 1) \tag{5}$$

It is to be noted that the apparent angle ϑ^* is included in the interval $[\vartheta_1, \vartheta_2]$. Figure 9 illustrates the behavior of the apparent Young angle according to the Cassie–Baxter relation (5).

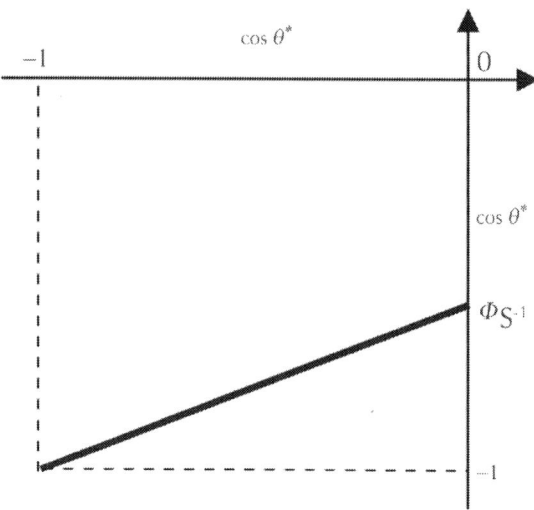

Figure 9: Apparent contact angle according to the angle of Young (Cassie–Baxter relation).

To summarize, a low roughness involves a Wenzel configuration while a strong roughness a Cassie–Baxter one. De Gennes showed that for a sinusoidal surface and a Young angle of 120°, the roughness from which appear air pockets is 1.75 [15]. Moreover, Bico et al. demonstrated that the Cassie–Baxter mode is thermodynamically stable for a given value threshold $\cos\theta_c$ [26]. The value of this angle can be determined when the drop is positioned in the Cassie–Baxter state, where its energy is minimized as compared to Wenzel mode. The variation of energy calculated from Eq. 4 must thus be weaker than that calculated from Eq. 2, from where:

$$\cos\theta_C = \frac{\phi_S - 1}{r - \phi_S} \tag{6}$$

This leads to a coexistence of the two modes, as described in Fig. 10:

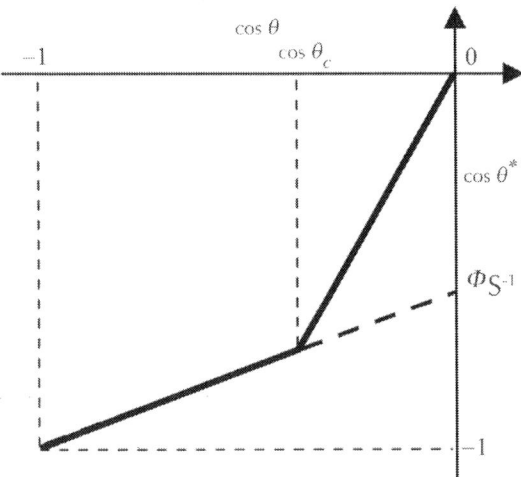

Figure 10: Coexistence of two superhydrophobic modes. With feeble hydrophobicity ($\cos\theta_c < \cos\theta < 0$), the apparent contact angle is theoretically given by the relation of Wenzel while for strong hydrophobicity ($\cos\theta < \cos\theta_c$), the apparent contact angle follows the relation of Cassie–Baxter. However, in practice, an average hydrophobicity generally involves a metastable configuration of Cassie–Baxter (dotted lines)

However, when a drop is deposited on a rough surface, a Cassie–Baxter regime occurs even when $\theta < \theta_c$ (for water, $\theta < 120°$) [27-29]. This state is metastable, i.e. by applying a pressure to the drop, for example, it is possible to reach the Wenzel regime: stable and displaying an important hysteresis [30]. This state is problematic, in particular in microfluidic microsystems where the displacement of a drop with a hysteresis of 100° is not easily realizable. An ideal configuration is the rolling ball or fakireffect i.e. the Cassie–Baxter state.

Neinhuis and Barthlott studied in detail the superhydrophobic properties of almost 200 plants, the famous lotus effect. In most cases, the surface comprises two different roughness scales: one is micrometric and the other one is nanometric.

The first assumptions on this double roughness were brought by Bico [31], Herminghaus [32] and many other teams [33, 34].

According to the work of Bico, this double roughness would avoid placing the drop in the Wenzel state; small asperities will trap air and as a consequence the drop will be in an intermediate configuration between Wenzel and Cassie–Baxter [21] (Fig. 11).

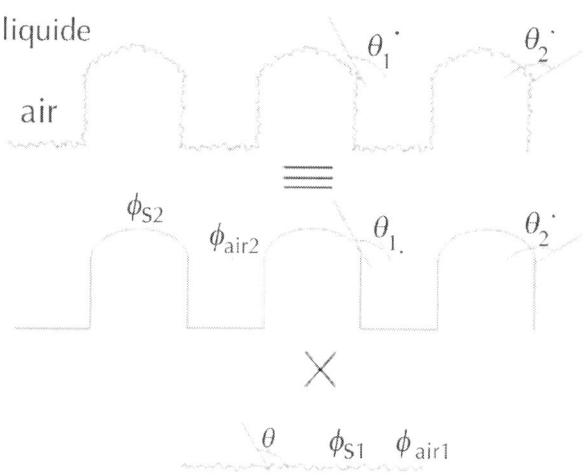

Figure 11. Apparent contact angle on a surface with two different roughness scales.

In the case of a double roughness, the equation of Cassie–Baxter becomes:

$$\cos \theta_2^* = \phi_{S1} \phi_{S2} \cos \theta - \phi_{S2} \phi_{A1} - \phi_{A2} \qquad (7)$$

With

$$\cos \theta_2^* = \phi_{S2} \cos \theta_1^* - \phi_{A2} \qquad (8)$$

And

$$\cos \theta_1^* = \phi_{S1} \cos \theta - \phi_{A1} \qquad (9)$$

where ϑ is the angle of Young, ϑ_1^*, Φ_{S1} and Φ_{A1} are respectively the angle, the solid fraction of surface and the fraction of air surface with nanometric roughness, and ϑ_2^*, Φ_{S2} and Φ_{A2} are respectively the angle, the solid fraction of surface and the fraction of air surface with micrometric roughness (Fig. 11). From Eq. 7, the double roughness amplifies the superhydrophobic surface property. If, for example, two roughnesses are homothetic, they have the same fraction of surface ΦS and the equation of Cassie–Baxter becomes:

$$\cos \theta^* = -1 + \phi_S^2(1 + \cos \theta) \qquad (10)$$

When $\Phi_S < 1$, $\cos\vartheta^*$ is smaller than in the case of a simple roughness, the contact angle increases.

PREPARATION OF SUPERHYDROPHOBIC SURFACES

From a technological point of view, there are currently several possibilities to mimic and prepare artificial superhydrophobic surfaces, including generating of rough surfaces coated with low surface energy molecules, roughening the surface of hydrophobic materials, and creating well-ordered structures using micromachining and etching methods. Some examples will be seen in the next part of this review.

WETTABILITY SWITCHING TECHNIQUES ON SUPERHYDROPHOBIC SURFACES

Carbon Nanotubes Anisotropic Structures

Carbon nanotubes (CNTs) are naturally hydrophilic. However, their wetting behavior is highly dependent on their arrangement and can vary from hydrophilic to hydrophobic and even superhydrophobic with in addition isotropic to anisotropic CA hysteresis. Two strategies have been developed to reach a stable superhydrophobic state. First a chemical modification of CNTs with a low surface energy compounds [mainly fluoropolymers like poly(tetrafluoroethylene) and silanes] leading to a CA as high as 171° with a roll off behavior, consistent with a quasi null hysteresis [35]. Second, hierarchical structures inspired by the 'lotus effect' were fabricated by CVD on a patterned quartz substrate, giving a CA of 166° with a CA hysteresis of 3°. Using an anisotropically rough surface, leading to an anisotropic CA, Jiang et al. have prepared a surface mimicking the rice leaf (a two dimensional anisotropy) showing that a droplet can roll along a determined direction [36]. As predicted by Jiang [37], three-dimensional anisotropic structured carbon nanotubes (ACNTs) can be designed with a gradient roughness distributed in a particular direction where the gradient wettability is predetermined and therefore the droplet may move spontaneously, driven by the wettability difference.

Mechanical

The first report on a switching wettability based on roughness modification by mechanism action was proposed by He [38]. The device consists of a thin polydimethylsiloxane (PDMS) membrane bound on a top of rough PDMS substrate. The switching was dynamically tuned from medium hydrophobic to superhydrophobic states by deflecting the membrane with a pneumatic method. The flat surface shows a contact angle of 114.6° while for the rough surface containing square pillars (26 × 24 µm² with a 25 µm height, giving rise to superhydrophobic classical droplet behavior), the CA is about 144.4°. Pneumatic actuation of the membrane leads to a CA difference of 29.8° (from flat to rough surface) (Fig. 12). The droplet displacement is only possible across the boundary of the patterned area: the droplet is gently deposited on the rough surface (i.e. after actuating the membrane) and moves to the flat one: receding angle on the rough surface is greater by 17° than the advancing angle on the flat surface. This contact angle difference can generate enough driving force to produce droplet motion from rough to flat surface. However, the droplet did not move for a reversible operation sequence (i.e. deposited on the flat surface then actuating the membrane). The authors explained the phenomenon by the formation of a wetted contact leading to a contact angle close to that on the flat surface. The driving force is not enough to cause droplet motion. A solution proposed by the authors to overcome this problem is to realize a double roughness of the surface in order to mimic superhydrophobic structures leaves.

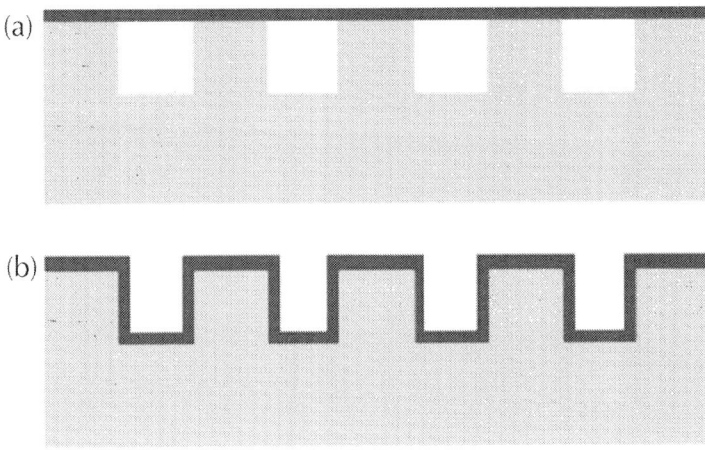

Figure 12: Concept of the thin membrane device: (a) with a flat surface, (b) pneumatic actuation leading to a rough surface.

Chen et al. [39] reported on the modification of surface wetting induced by morphology change (SWIM). A conductive metal/polymer composite membrane, supporting hydrophobic microposts of various heights, is sustained by negative photoresist spacers (Fig. 13). Before applying an electrical potential (initial state) a droplet is bolstered on the higher microposts with a contact angle of 152°. When a voltage (250 V) is applied between the conductive polymer membrane and the bottom addressable electrodes (actuated state), the membrane is bent (10 μm vertical displacement) due to the electrostatic force, and the highest microposts are lowered down. The droplet sticks to the lower posts and the contact angle decreases to 131°. Unfortunately, the authors did not indicate clearly the reversibility of the phenomenon, and did not precise the hysteresis observed for these surfaces. Nonetheless, an advantage of this mechanical device is a free electric interference mechanism compared to electrowetting and prevents the surface from nonspecific adsorption of proteins on the hydrophobic layer.

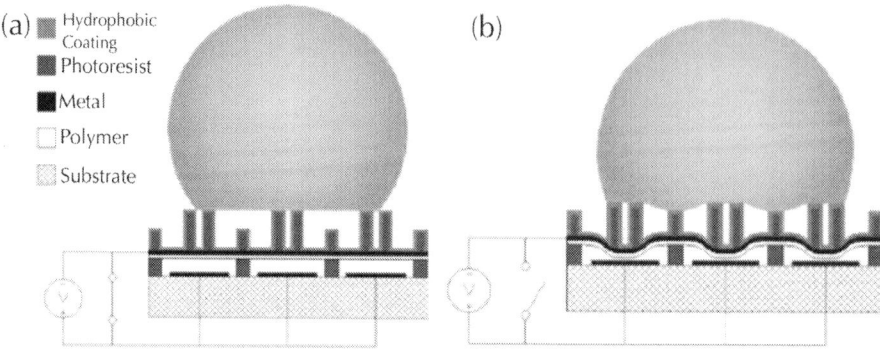

Figure 13: The operation concept of SWIM: (a) at initial state, the droplet merely contacts the higher posts and (b) at actuated state, the droplet will contact with both the higher and lower posts. Reprinted with permission from [39]. Copyright 2007 Institute of Physics.

Zhang et al. [40] described a method to generate reversible wettability upon switching between superhydrophobicity and superhydrophilicity by biaxially extending and unloading an elastic polyamide film with triangular net-like structure composed of fibers of about 20 μm in diameter. The average side of the triangle of the net-like structure is around 200 μm before biaxial extending (with a CA of 151.2°) and 450 μm after extension (with a CA of 0 ± 1.2°) (Fig. 14). The mechanical actuation presented in this part consists mostly in increasing the liquid/solid surface (resulting in the modification of the apparent contact angle) rather than modifying directly the surface wetting properties.

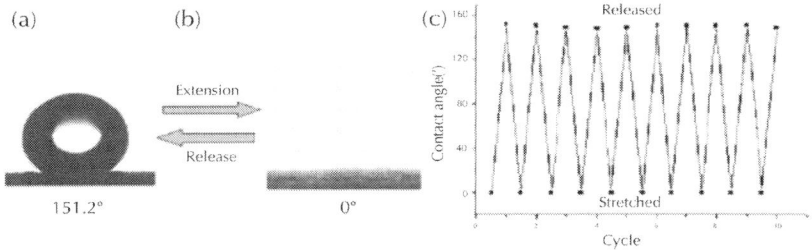

Figure 14: Switching between superhydrophobicity and superhydrophilicity of an elastic polyamide film with a triangular net-like structure. (a) Before biaxial or after unloading, the CA is about 151°. (b) When the film was extended, the CA is around 0° (i.e. reversible superhydrophobic/superhydrophilic transition of the films by biaxial extension and unloading). Reprinted with permission from [40]. Copyright Wiley-VCH Verlag GmbH & Co. KGaA.

Magnetic

A superhydrophobic surface was used for reversibly oriented transport of superparamagnetic microliter-sized liquid droplets with no lost volume in alternating magnetic fields. The surface consists of an aligned polystyrene (PS) nanotube layer prepared via a simple porous alumina membrane template covering method [41]. This surface displays a superhydrophobic behavior (CA of about 160°) with a strong adhesion force to water, as compared to traditional superhydrophobic surfaces. Instead of estimating the hysteresis of the surface, the authors measured the adhesive force. According to their results, adhesive forces of the surfaces were 10 times higher than that of a surface displaying a water CA hysteresis of 5°, proving the Wenzel state of the droplet. They used a super paramagnetic microdroplet (for an intensity of external magnetic field ranging from 0.3 to 0.5 T) placed on an ordinary superhydrophobic surface (CA of 160°), separated from the PS surface with 2 mm in height [42].

When the upper magnet was applied, the microdroplets were magnetized, fly upward and stick to the PS surface due to its strong hysteresis. On the other hand, when the magnetic force was reversed, the microdroplet fell down onto the initial surface. The principal key point of this application is that the reversible transport is made without any lost of liquid.

Chemical

A two-level structured surface (SAS) of polymer has been synthesized by Zhou and Huch [43]. The first level of roughness (~1 μm) was obtained by plasma etching of a rough polymer film (PTFE). Then surface hydroxyl and amino functional groups have been introduced by plasma treatment in order to form a grafted mixed brush consisting of two carboxyl-terminated incompatible polymers PSF-COOH and P2VP-COOH. After exposure to toluene, an advancing contact angle of 160° was measured with no angle hysteresis (rolling ball state). After immersion of the sample in an acid (pH 3) bath for several minutes and its subsequent drying, a drop of water spreads on the surface. The authors clearly indicate that the superhydrophobic state is time dependant. Up to a few minutes after exposure to toluene, the surface was superhydrophobic with quasi null hysteresis, while the hysteresis increases dramatically with time due to the slow switching of the surface composition to a more hydrophilic state.

Temperature

The first demonstration on thermal reversible switching behavior between superhydrophilicity and superhydrophobicity was reported by Sun et al. [44]. They used a thermo responsive polymer poly(N-isopropylacrylamide) (PNIPAAm) that exhibit, when deposited on a flat surface, a CA modification from 63.5° for a temperature of 25 °C (hydrophilic state due to the formation of intermolecular hydrogen bonding between PNIPAAm chains and water molecules) to 93.2° at 40 °C (hydrophobic state due to intramolecular hydrogen bonding between C=O and N–H groups of the PNIPAAm chains). The roughness effect on the wetting properties was further investigated by depositing the polymer on rough surfaces (obtained by a laser cutter on a silicon wafer) formed of a regular array of square silicon microconvexes (grooves of about 6 µm width, 5 µm depth and spacing from 31 to 6 µm). The obtained results clearly show that when the substrate is sufficiently rough (i.e. when groove spacing is smaller or equal to 6 µm), the thermally responsive switching between superhydrophilicity and superhydrophobicity can be realized: from a CA of 0° below T = 29 °C to 149.5° above 40 °C, indicating that a combination of the change in surface chemistry and surface roughness can enhance stimuli-responsive wettability.

Fu et al. [45] have developed a slightly different approach based on porous anodic aluminum oxide (AAO) template with nominal pore sizes from 20 to 200 nm. The grafting of PNIPAAm on the template was obtained by surface-initiated atom transfer radical polymerization (ATRP) leading to a reproducible and uniform brush film (15 nm thick) on the textured surface. According to the authors, the macroscopic wettability is not due only to the change of the polymer hydrophobicity, but also to the nanoscopic topography of the surface associated with expansion and contraction of the grafted polymer. Nonetheless, these surfaces led to a maximum contact angle of 158° at 40 °C (for 200 nm pore size) starting from a CA of 38° at 25 °C, comparable to the contact angles reported by Sun et al.[44].

Dual Temperature/pH

Xia et al. [46] have prepared a dual-responsive surface (both temperature and pH) that reversibly switches between superhydrophilic and superhydrophobic. In addition, the lower critical solubility temperature (LCST) of the copolymer is tunable with increasing the pH. The copolymer thin film is a poly(N-isopropyl acrylamide-co-acrylic acid) [p-(NIPAAm-co-AAC] deposited on a roughly etched silicon substrate composed of patterned square pillars (20 µm high, 12 µm long, and 6 µm spacing between the silicon pillars). For a pH = 7, identical behavior, from superhydrophilic to superhydrophobic was obtained, as compared to classical PNIPAAm discussed above. However, for pH values of 2 and 11, the surfaces are superhydrophobic and superhydrophilic, respectively, whatever the temperature (Fig. 15). Another point is that, as compared to previously related reports on thermally responsive materials, the film can be hydrophobic at low temperature and hydrophilic at high temperature. These phenomena can be linked to the reversible change in hydrogen bonding between the two components (NIPAAm and AAc). It is to be noted that the transformation from superhydrophobic to superhydrophilic takes several minutes (time for a single cycle).

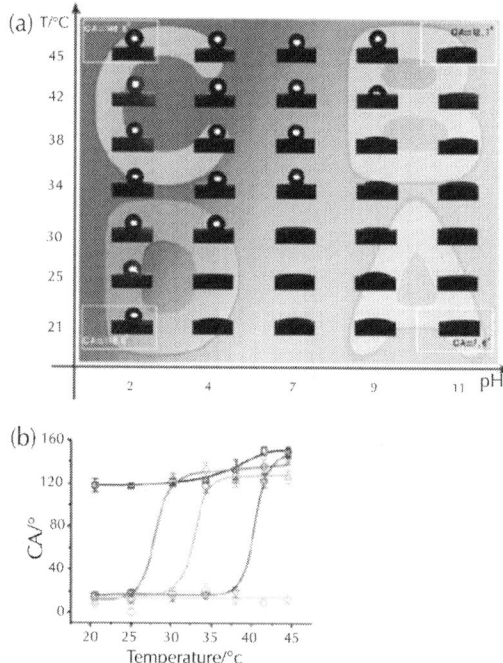

Figure 15: (a) When the pH and/or temperature is varied the CAs reversibly change. (b) Temperature and pH dependence of water CAs for P(NIPAAm-co-AAc) thin films. Water CAs change at different temperatures for a modified substrate at pH values of 2 (□), 4 (○), 7 (▲), 9 (▼) and 11 (◊), respectively. Reprinted with permission from [46]. Copyright Wiley-VCH Verlag GmbH & Co. KGaA.

Optical

The first example showing that the wetting characteristics of polymer surfaces doped with photochromic spiropyran molecules can be tuned when irradiated with laser beams of properly chosen photon energy was reported by Athanassiou et al. [47]. The hydrophilicity was enhanced upon UV laser irradiation since the embedded nonpolar spiropyran molecules were converted to their polar merocyanine isomers. The process is reversed upon green laser irradiation. To enhance the hydrophobicity of the system, the photochromic polymeric surfaces were structured using soft lithography. Water droplets on the patterned features interact with air trapped in the microcavities, creating superhydrophobic air–water contact areas. Furthermore, the light-induced wettability variations of the structured surfaces are enhanced by a factor of 3 compared to those on flat surfaces. This significant enhancement is attributed to the photoinduced reversible volume changes of the imprinted gratings, which additionally contribute to the wettability changes induced by the light. In this work, it was demonstrated how surface chemistry and structure can be combined to influence the wetting behavior of polymeric surfaces. However, the contact angle values after the UV and green light irradiation are limited to the first two UV–green irradiation cycles. The aging and degradation of the system upon multiple irradiation cycles is the major drawback of such a polymeric system.

On the other hand, Lim et al. [48] have reported a photoswitchable nanoporous multilayer film with wettability that can be reversibly switched from superhydrophobicity to superhydrophilicity under UV/visible irradiation. They used a combination of surface roughness and a photoresponsive molecular switching of fluorinated azobenzene molecule (7-[(trifluoromethoxyphenylazo)phenoxy] pentanoic acid (CF3AZO)). The surface roughness was obtained using a layer-by-layer deposition technique of poly(allylamine hydrochloride) (PAH)), which is a polyelectrolyte, and SiO_2 nanoparticles as polycation and polyanion, respectively giving a porous organic–inorganic hybrid multilayer films on silicon surface. In their study, the surface roughness can be precisely tuned by controlling the number of PAH/SiO$_2$ NPs bilayers. The film was further modified by 3-(aminopropyl)triethoxysilane to introduce amino groups serving as binding sites for the photoswitchable moiety. The wettability is dependent on the change of the dipole moment of the azobenzene molecules upon trans to cis photoisomerization (Fig. 16). For example, in the trans state, the azobenzene molecules exhibit the fluorinated moiety leading to a lower surface energy. The trans-to-cis isomerization of azobenzene is induced by UV light irradiation and leads to a large increase in the dipole moment of these molecules demolishing the chain packing in the azobenzene monolayer and a lower contact angle (the fluorinated moiety was not anymore exhibited). By this technique, the contact angle can be controlled by adjusting the number of multielectrolyte layers. A contact angle of 152° and a hysteresis below 5° was obtained for 9 bilayers with a little degradation after many cycles. They showed that patterning surface with hydrophilic and superhydrophilic zones can be easily achieved by using selective UV irradiation through an aluminum mask.

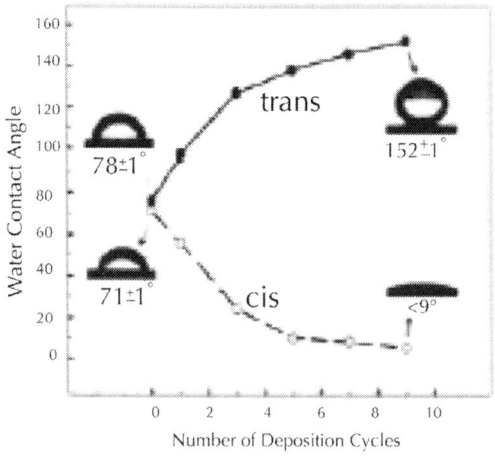

Figure 16: The relationship between the number of deposition cycles and the water contact angles: water droplet profiles on the smooth substrate (dotted arrows) and on the organic/inorganic multilayer film (solid ar-

rows) after UV/visible irradiation. Reprinted with permission from [48]. Copyright 2006 American Chemical Society.

The photoswitchable wettability of aligned SnO_2 nanorod films was demonstrated by Zhu et al. [49]. The SnO_2 nanorod films were prepared in two steps. First, SnO_2 seeds were spin-coated on a silicon substrate and then immersed in 50 mL aqueous solution of $SnCl_4 \cdot 5H_2O$ in the presence of urea and HCl in a closed bottle. The mixture was heated at 95 °C for 2 days to yield SnO_2 nanorod films. The resulting films were rinsed thoroughly with deionized water, dried at room temperature and stored in the dark for several weeks. The as-prepared SnO_2 nanorod films showed superhydrophobic behavior (contact angle of 154°), as compared to 20° displayed by a smooth SnO_2 surface. SnO_2 nanorod films changed to superhydrophilic state (0°) just by exposition to UV irradiation (254 nm) for 2 h. Then, the wettability goes back to its initial superhydrophobic state by keeping the films in the dark for a given time (4 weeks) [49] (Fig. 17). The switchable wettability was explained by the generation of hole-electron pairs after UV-irradiation on the surface of the SnO_2 nanorods reacting with lattice oxygen to form surface oxygen vacancies. The defective sites are kinetically more favorable for hydroxyl adsorption than oxygen adsorption, leading to the superhydrophilic state. During dark storage, hydroxyls adsorbed on the defective sites can be gradually replaced by oxygen in the air, because oxygen adsorption is thermodynamically more stable and lead to superhydrophobic state. Feng et al. showed similar switchable wettability properties for ZnO nanorod films [50]. In these cases, the reversible switching between superhydrophilicity and superhydrophobicity is related to the cooperation of the surface chemical composition and the surface roughness. The former provides a photosensitive surface, which can be switched between hydrophilicity and hydrophobicity, and the latter further enhances these properties.

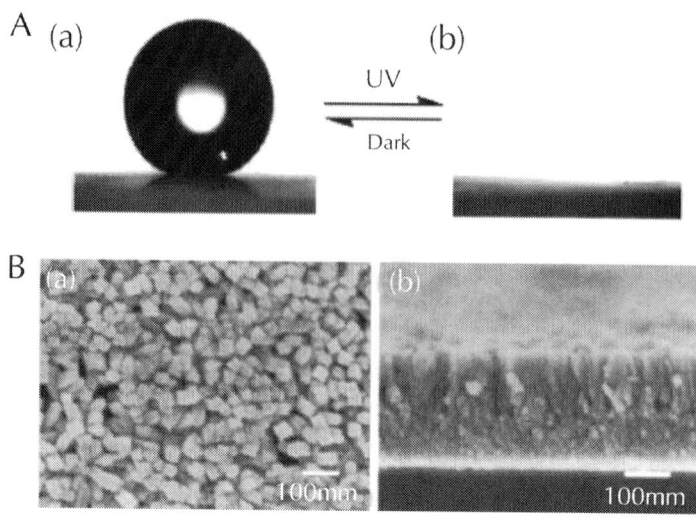

Figure 17: (A) Water droplet shapes on as-prepared SnO_2 nanorod films (a) before and (b) after UV-irradiation; (B) (a) and (b) are the top and cross-sectional FE-SEM images of the as-prepared SnO_2 nanorod films, respectively. Reprinted with permission from [49]. Copyright 2007 Royal Society of Chemistry.

By using titania nanoparticles, a patterning and tuning method of microchannel surface wettability was developed for microfluidic control [51]. Titania modification of a microchannel was achieved by introduction of titania solution inside pyrex microchannel providing a nanometer-sized surface roughness. Subsequent hydrophobic treatment with ODS (octadecyl dichlorosilane) gavelled to superhydrophobic surface (contact angle of 150°). Photocatalytic decomposition of the coated hydrophobic molecules was used to pattern the surface wettability, which was tuned from superhydrophobic to superhydrophilic under controlled photoirradiation (Fig. 18). Irradiation for 60 min gave a superhydrophilic surface (9°). This wettability changes were explained by the small number of ODS molecules covering the titania surface caused by photocatalytic decomposition of ODS. Furthermore, a four-step wettability based Laplace valves working as passive stop valves were prepared by using the patterned and tuned surface. As a demonstration, a batch operation system consisting of two sub-nL dispensers and a reaction chamber was constructed. Fundamental liquid manipulations required for the batch operation were successfully conducted, including liquid measurement (390 and 770 pL), transportation, injection into the chamber, and retention in the chamber. To verify the quantitative operation, the system was applied to a fluorescence quenching experiment as an example of volumetric analyses. The method provides flexible patterning in a wide range of tuned wettability surfaces in microchannels even after channel fabrication and it can be applied to various two- or multi-phase microfluidic systems.

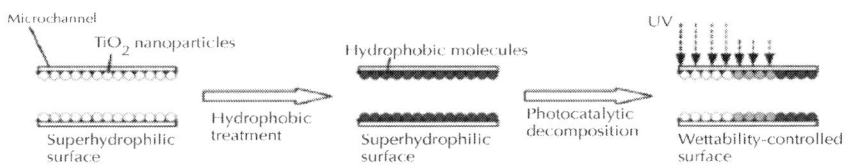

Figure 18: Photocatalytic patterning and tuning of surface wettability by photoirradiation of modified titania nanoparticles. Reprinted with permission from [51]. Copyright 2007 Royal Society of Chemistry.

Another example of titanium-based material was described by Balaur et al. [52]. They used self-organized TiO_2 nanotube layers grown on Ti by electrochemical anodization. The as-prepared TiO_2 nanotubes displayed a superhydrophilic wetting behavior. When modified with organic molecules, such as octadecylsilane or octadecylphosphonic acid layers, the surfaces showed a superhydrophobic behavior. They have demonstrated how the tubular geometry of the TiO_2 layers combined with an irreversible UV induced decomposition of the organic monolayers can be used to adjust the surface wetting properties to any desired degree from super-hydrophobic to superhydrophilic (Fig. 19).

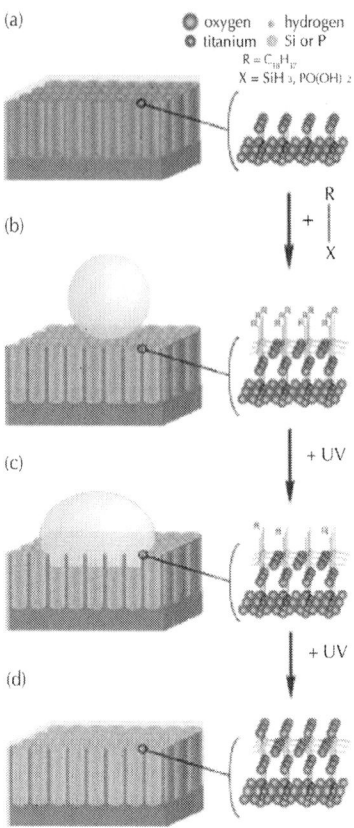

Figure 19: Schematic illustration of the process used to adjust contact angles. The scheme shows the different stages of the wetting behavior: (a) the nanotube surface; (b) superhydrophobicity after hydrophobic modification; (c) chain scission of the organic layer triggered by UV light and (d) leading finally to complete wetting. Reprinted with permission from [52]. Copyright 2005 Elsevier.

Nanowires can also be used for the preparation of superhydrophobic surfaces with a tunable wettability. Coffinier et al. presented a simple method for producing superhydrophobic surfaces based on chemical modification of silicon oxide nanowires [53]. Nanowires with an average mean diameter in the range of 20–150 nm and 15–20 μm in length were obtained by the so-called solid–liquid–solid (SLS) mechanism at 1,100 °C under N_2 flow during 60 min. The porous nature and the high roughness of the resulting surfaces were confirmed by AFM imaging. After cleaning, the silicon nanowires have been modified by PFTS (perfluorodecyl trichlorosilane), resulting in a superhydrophobic surface with a contact angle of 152°, which is much higher than that of a smooth Si/SiO$_2$ surface modified with the same silane (109°) (Fig. 20). The contact angle of the unmodified surface was closed to 0°, as expected for a surface terminated with polar hydroxyl (OH) groups. The surface wettability can be irreversibly tuned by controlling the UV-irradiation time, resulting in a partial or complete removal of the organic layer. The chemical modification and degradation of the organic layer was followed by XPS analysis.

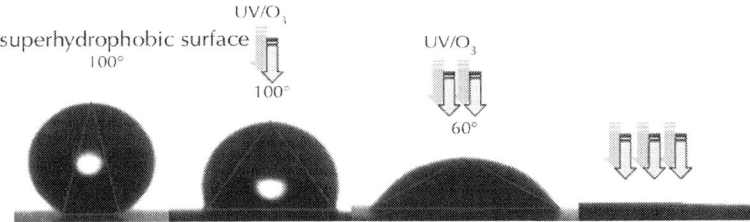

Figure 20: Control of wettability of PFTS-terminated silicon oxide nanowires as a function of exposition time to UV-irradiation.

EWOD

Theory and History

Lippmann showed, during his thesis on electrocapillarity in 1875 [54], that the application of a voltage between an electrolyte and a drop of mercury immersed in this one involved the creation of a double electric layer (EDL, Electric Double Layer) at the interface. The electrowetting principle consists, starting from the electrocapillarity phenomenon, to modify the shape of a liquid droplet placed on a surface during the application of a voltage (Fig. 21). Since the majority of the liquids used in Lab-on-Chip devices are conductive, the idea developed by Berge was to isolate the drop from the substrate using a dielectric layer in order to avoid any phenomenon of electrolysis [55]. This development is known as ElectroWetting On Dielectric (EWOD).

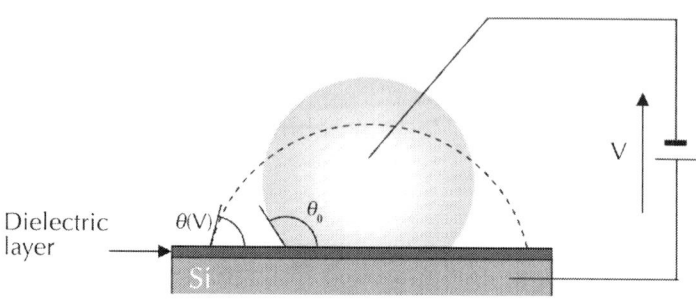

Figure 21: EWOD principle. Under applied voltage, the drop spreads out on the surface.

The system can be seen like a variable capacitor [56]. The energy stored in this capacitor according to a direction perpendicular to the plan, noted W(x), is expressed by:

$$W(x) = \frac{1}{2}C(x)V^2 = \frac{\varepsilon_0 \varepsilon_r}{2e} xV^2 \tag{11}$$

where ε_r is the permittivity of the dielectric layer, ε_0, the electric permittivity of the vacuum, x, the length of the capacitor and E, its thickness. By applying the principle of virtual work, the force per transverse unit of length is deduced:

$$F_m = \frac{\partial W(x)}{\partial x} = \frac{\varepsilon_0 \varepsilon_r}{2e} V^2 \tag{12}$$

This force, acting on the three phase contact line, can be inserted in the equation of Young (1):

$$\gamma_{LS} = \gamma_{SG} - \gamma \cos \theta(V) + \frac{\varepsilon_0 \varepsilon_r}{2e} V^2 \tag{13}$$

Equation 1 leads then to the equation of Young–Lippmann established by Bruno Berge in 1993:

$$\cos \theta(V) = \cos \theta_0 + \frac{\varepsilon_0 \varepsilon_r V^2}{2\gamma e} \qquad (14)$$

Although, Young–Laplace pressure works in prediction of droplet shape modification by EWOD, different theories have been proposed to explain the real nature of the movement. Historically, electrowetting was explained by the variation of interfacial energies: the increase of the voltage leads to a solid–liquid interfacial energy diminution [57]. More recently, it has been proved that EWOD can be interpreted as an electromechanical effect: pressure exerted by electrical field on the drop surface acts on the contact line [58-60]. While this last view seems to be the correct one, both of them predict the same contact angle variation [61, 62].

Furthermore, according to Eq. 14, it is theoretically possible to obtain a total wetting of the drop by increasing the applied voltage. However, a saturation of the contact angle is observed starting from a certain voltage. The literature brings many assumptions for the comprehension of this saturation like an increase in the electric field to the level of the three phase contact line due to pick effect [63], trapping of charges in or on the dielectric layer [64,65], ionization of air on the level of the triple line[66], leakage on the dielectric layer, [67]. Nevertheless, while reasons for this saturation are not clearly established by the scientific community, in practice the maximum tension V_{max} to be applied for electrowetting is always observed.

Optical Applications of EWOD

This part of the review, which is not exhaustive deals with the potential applications of the EWOD technique. For more detailed state of the art as well from the theoretical point of view, refer to recent reviews by Mugele and Baret [68] (which in addition contains an English version of the thesis of Lippmann on electrocapillarity), and by Fair [69]. Berge was the first to bring a microsystem based on EWOD to maturation at the industrial level with liquid lenses [70]. The principle is simple and is schematically represented in Fig. 22. Oil and water drops are trapped between two transparent substrates. The spacing between the two substrates is ensured by metal electrodes. At V = 0 V, the drops form a certain contact angle with the surface. The formed meniscus thus has a defined radius of curvature, and optical rays are divergent (Fig. 22a). Upon application of a tension of ~60 V, the contact angle changes, the radius of curvature is modified, the luminous rays are focused (Fig. 22b).

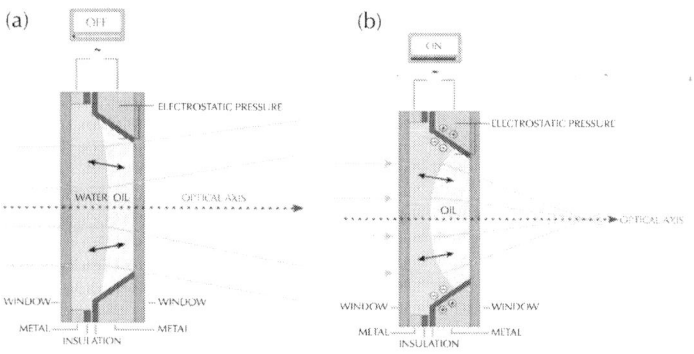

Figure 22: Principle of Varioptic liquid lenses operation based on EWOD principle: (a) the tension is cut off, the rays are divergent, (b) the tension is applied, the rays are focalized [71]. Reprinted with permission from Varioptic.

Figure 23 exhibits two models of lenses. The market aims primarily that of mobile telephony. Recently, Varioptic commercialized its first autofocus module, in partnership with Sunny Optics (China). These lenses have several advantages, as compared to the traditional lenses. First of all, the absence of moving parts allows a better integration. The weak voltage required for actuation allows the introduction of autofocus modules into the mobile telephones. Lastly, the lens has a perfect surface since it is about the interface between two liquids with a price divided by 10.

Wettability Switching Techniques on Superhydrophobic Surfaces 117

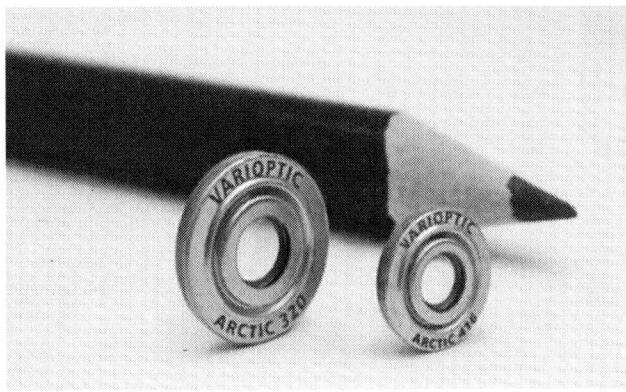

Figure 23: Two models of lenses developed by Varioptic. Reprinted with permission from Varioptic.

Several teams work on the development of such lenses. The principal stake is the reduction of the tension, necessary to the operation of the lens. The team of Heikenfeld at the University of Cincinnati developed a concept of optical prism by obtaining flat meniscuses for a drop taken between two substrates [72]. By applying a specific tension to each substrate, it is possible to vary the orientation of the prism (Fig. 24) [73].

Figure 24: Response of the prisms according to the applied voltage to each substrate. Reprinted with permission from [73]. Copyright 2006 Optical Society of America.

EWOD allows also visualizing images thanks to screens containing liquid pixels controlled by electrowetting. A spin-off of Philips, Liquavista [74], develops color screens based on electrowetting. The market aimed with such screens is always that of mobile telephony. The principle is similar to that of the Varioptic lenses. Each pixel consists of a water drop, which lets pass the light, and of an oil drop, opaque or of color. If no voltage is applied, the oil drop spreads out, the light does not go through (or the pixel is colored). On the other hand when a voltage is applied, the water takes the place of the oil, resulting in a white pixel [75]. A general diagram of a monochromic and fluorescent pixel is presented in Fig. 25[76]. In the case of the pixel developed at the University of Cincinnati by Heikenfeld, the principle is the reverse. The pixel is fluorescent if no voltage is applied (fluorescent oil for $\lambda = 405$ nm). Once a voltage is applied, the water takes the place of the oil and the light is completely reflected, the pixel is extinct.

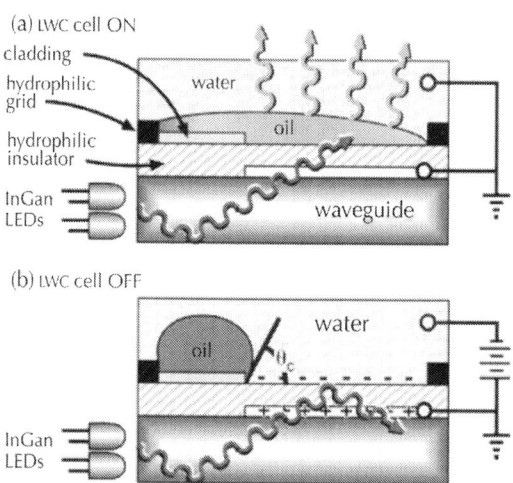

Figure 25: General diagram of the liquid pixel for fluorescent screen produced at the University of Cincinnati. Reprinted with permission from[76]. Copyright 2005 American Institute of Physics.

EWOD for Microdroplets Displacement

In order to displace microdroplets and to realize microfluidic basic operations (merging, creating droplets), the EWOD system needs to have two plans: a base composed of electrodes for displacement and a counter-electrode (instead of a needle). A general diagram of the two plans microsystem is shown in Fig. 26. Initially, no voltage is applied between the electrodes and the counter-electrode, and whatever the place where the drop is placed, the contact angle is the contact angle of the drop θ_0. When a voltage is applied on an electrode under the drop, the contact angle of the three phase contact line in contact with this electrode decreases to reach a value θ_d and thus the radius of curvature R_d of the meniscus increases. The contact angle on the rest of the substrate is always the contact angle to balance θ_0 and the associated radius of curvature R_0 is lower than the radius of curvature R_d.

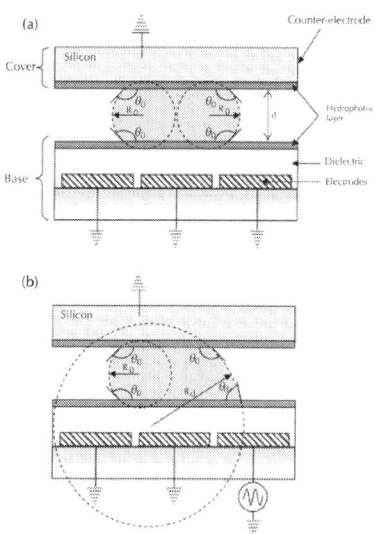

Figure 26: General set up of an EWOD microsystem for the displacement of microdroplets: (a) no voltage is applied to the electrode, (b) a voltage is applied to the electrode of right-hand side.

According to the Laplace law, the meniscus curvature radius change involves a difference in pressure within the drop [77]. This pressure difference is given by:

$$\Delta P = P_g - P_d \tag{15}$$

where P_g is the pressure on the left side in the drop whereas P_d is the pressure on the right side. These two values are determined by the following expressions:

$$P_g - P_a = \gamma \left(\frac{1}{R_0} + \frac{1}{R} \right) \quad (16)$$

$$P_d - P_a = \gamma \left(\frac{1}{R_d} + \frac{1}{R} \right) \quad (17)$$

where P_a is the atmospheric pressure, R the ray of the drop in the transverse direction, R_0, the radius of curvature of the left meniscus and R_d, the radius of curvature of the right meniscus. Thus,

$$\Delta P = \gamma \left(\frac{1}{R_0} - \frac{1}{R_d} \right) > 0 \quad (18)$$

The pressure within the drop is stronger on the left than on the right, the drop moves on the electrode of right-hand side. So with:

$$R_0 = -\frac{d}{2 \cos \theta_0}$$

$$R_d = -\frac{d}{\cos \theta_0 + \cos \theta_d}$$

We found:

$$\Delta P = \gamma \frac{\cos \theta_d - \cos \theta_0}{d} \quad (19)$$

Starting from Eq. 19, the driving force F_m, which allows displacement, can be deduced (per unit of length):

$$F_m = \gamma (\cos \theta_d - \cos \theta_0) \quad (20)$$

The force F_m drives the drop on the electrode under applied voltage. Until now, all the calculations were applied for perfect surfaces. However, certain forces such as hysteresis or viscous forces can hinder the displacement of the drop. Fouillet showed by digital simulation that the movement of the drop is related to the interfacial forces and not to the viscous forces [78]. Concretely, it is necessary that the driving force is higher than the force of hysteresis in order to obtain a displacement of the drop. Within the framework of real surfaces, it is thus necessary that the driving force is higher than the force of hysteresis.

Lab-on-Chip Applications

Although the industrial applications of the EWOD are in the field of optics, several groups are also interested in the possible applications in biotechnology. For this purpose, it is necessary to displace biological liquids and to realize microfluidic elementary operations for the development of Lab-on-chip, LoC. The LoC based on EWOD were initiated by Pollack et al., from the Duke University [79,80]. By carrying out a series of electrodes, it is possible to move by EWOD effect the drop from one electrode to its neighbor by successive polarization. In this case, the electrodes are made of chromium; the dielectric is parylene C (700 nm thick) covered with Teflon (200 nm thick). The counter-electrode is a covered blade of glass ITO and Teflon. The gap between the two substrates is 300 µm for electrodes of 1.5 mm^2. The displacement of drops of KCl (100 mM) was carried out under a tension of 120 V$_{DC}$. In 2004, the same team has developed a Lab-on-Chip based on EWOD allowing the determination of the concentration of glucose in a drop of plasma, serum, urine and saliva [81]. The detection scheme was based on the change of absorbance of the sample mixture/reactive versus time.

Other Lab-on-Chip devices have been realized by research teams from the University of Los Angeles, USA and CEA-Grenoble, France. Kim and Garrell from the University of Los Angeles (UCLA) developed a device offering the possibility to carry out several operations, including MALDI mass spectrometry analysis [82]. A microsystem comprised of different zones for sample purification and MALDI analysis is illustrated in Fig. 27. The method consists in moving a drop of biological liquid containing peptides and other impurities (urea, salts) by electrowetting on a hydrophobic Teflon pad. Peptides are adsorbed on the surface by hydrophobic/hydrophobic interactions. A water drop, moved by EWOD, dissolves the impurities mixed with peptides. Finally, a drop of matrix is brought on the pad and the microsystem is introduced into a MALDI mass spectrometer. At the same period, similar microsystems have been developed and patented within the framework of contract BIOCHIPLAB [83].

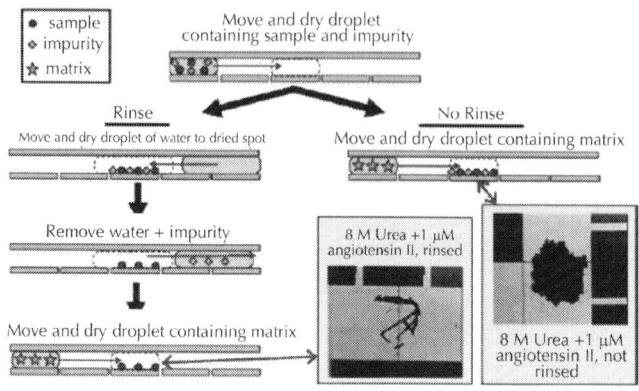

Figure 27: Lab-on-Chip principle for MALDI mass spectrometry analysis developed by Kim and Garrell. Reprinted with permission from [82]. Copyright 2005 American Chemical Society.

DISCUSSION

The hysteresis effect and the saturation phenomenon limit the interval of tension to be used for EWOD. Concretely, the voltage allowing displacement must lie between V_{min} (related to hysteresis) and V_{max} (related to saturation). The microsystems have most of the time vocation to be embarked. It is thus necessary to reduce the tensions of actuation. One of the solutions is the development of 1 plan microsystems, i.e. without counter-electrode [84]. In this case, the force related to hysteresis is only reduced by a factor $\sqrt{2}$, which is still not very practical in an embarked system. Moreover, such microsystems are definitely more sensitive to evaporation and do not allow microfluidic operations like drop scission. Another solution consists to reduce the thickness of the dielectric layer or to increase the permittivity of this one. However, a reduction in the dielectric layer involves an increase in the electric field. Under a certain thickness, the electric field is higher than the dielectric rigidity and involves a breakdown of the layer. There is thus a limit in the reduction of tension. The increase in the permittivity of the dielectric layer is limited by the weak permittivity of the hydrophobic layer. Thus, there is a breakdown even when a voltage of only few volts was applied [63].

The last possibility is the reduction of the hysteresis by using superhydrophobic surfaces (with hysteresis lower than Teflon).

Nonreversible Electrowetting on Superhydrophobic Surfaces

Up to date, all the teams working on electrowetting on superhydrophobic surfaces encountered the same problem: a drop wedged in a nanostructure does not go up, leading to an irreversible EWOD effect. Several groups have tried for the last few years to obtain a reversible electrowetting phenomenon, but unsuccessfully. Krupenkin [85] from the Bell Lab (USA) is one of the precursors in this field. The surfaces employed in the study are composed of silicon pillars, engraved through a mask carried out by electronic lithography ('fakir carpet' geometry). The electric insulation is ensured through oxidation of the surface. Upon applying a voltage, a total damping of the drop on the surface was observed, as shown in Fig. 28. Unfortunately, this phenomenon proves to be irreversible.

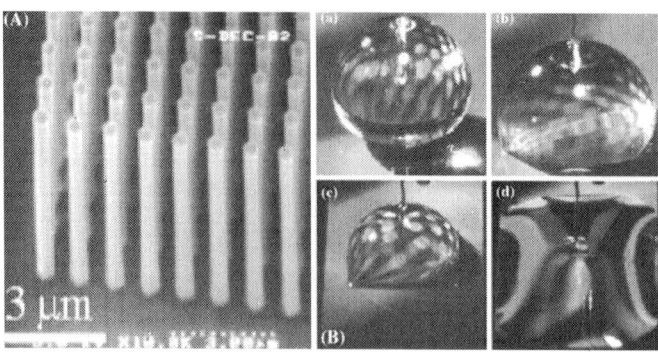

Figure 28: (A) SEM image of the silicon nanostructure used for electrowetting, (B) total wetting by electrowetting of a drop of cyclopentanol on an e-beam nanostructured surface: (a) no tension is applied, (d) total wetting under application of a tension (50 V). Reprinted with permission from [85]. Copyright 2004 American Chemical Society.

The same group brings in 2005 a first solution for the reversible wetting on such surfaces [86]. A very short electrical current impulse applied to the substrate leads to the surface heating. The temperature can then reach 240 °C, causing liquid boiling and droplet expelling from the surface. Even though this technique is easy to implement, it is hard to imagine such an integrated system within a Lab-on-Chip. The heating would cause significant damage to biological material within the drop. Moreover, this expulsion creates satellite droplets.

Other teams worked on electrowetting on textured surfaces by using various materials, like SU-8 [87] or carbon nanotubes (CNT) [88]. In the first case, the reversibility is not total. The angle decreases from 152° to 90° under 130 V and returns back to 114° when the tension is cut off. In the second case (CNT), no reversibility is observed. A solution allowing the reversibility is to modify the ambient conditions. Indeed, the irreversibility is observed when the ambient condition is air. By replacing air by a hydrophobic medium, like oil (dodecane), it is possible to obtain reversibility as shown in the Fig.29. The angle decreases from 160° to 120° (100° in air) when a tension was applied and returns back to 160° after tension cut off (Fig. 29).

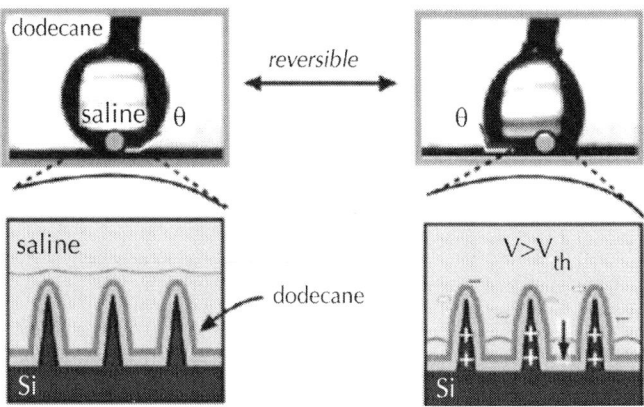

Figure 29: Reversibility of EWOD phenomenon on superhydrophobic surface by immersion of the water drop in dodecane. Reprinted with the permission from [88]. Copyright 2006 American Chemical Society.

It is interesting to notice that an oil environment prevents the Wenzel effect. However, the question of the applicability of such a surface is not clearly explained since a water drop in an oil environment has already a very high contact angle [89], even on a planar surface.

A fast calculation makes it possible to determine the angle of a water droplet on Teflon in an oil environment starting from the equation of Young:

$$\gamma_{ES} = \gamma_{SH} - \gamma_{EH} \cos \theta_0 \quad (21)$$

With

$$\gamma_{ES} = 47 \text{ mN m}^{-1}$$

$$\gamma_{SH} = 2 \text{ mN m}^{-1}$$

$$\gamma_{EH} = 50 \text{ mN m}^{-1}$$

We found:

$$\theta_E = 154°$$

Thus a planar surface allows at the same time a total wetting but also a complete reversibility.

Recently, Heikenfeld has reported electrowetting applied to textiles [90]. Two electrowetting textiles were prepared. The first one is made of a polyethylene naphthalate (PEN) film containing holes coated with Al (50 nm) (conductive layer). The second one was fabricated from wood microfibers on which a polymer (PEDOT-PSS and PEI) was deposited to render it electrically conductive. In both case parylene C (1 μm) and a fluoropolymer solution were used to insure a hydrophobic dielectric surface coating. The textile surfaces investigated are highly irregular and their electrowetting behavior was predicted, in first approximation by Cassie Baxter equation. For both textiles, irreversible electrowetting was observed with a contact angle varying from 120° to 70° in air. Here again, reversible electrowetting occurs in an oil environment.

Reversible Electrowetting on Superhydrophobic Surfaces

Our group has developed a different strategy to achieve electrowetting on superhydrophobic surfaces using a very heterogeneous surface composed of silicon nanowires coated with a fluoropolymer C_4F_8 [91]. The SiNWs were grown on Si substrate using the vapor–liquid–solid (VLS) mechanism and electrically insulated with 300 nm of SiO_2, First, a thin film of gold (4 nm thick) was evaporated on the substrate and then exposed to silane gas at different pressures at 500 °C for a given time. According to time and pressure of growth, eight surfaces were realized where the nanowires length varied from 1 μm (10 min, 0.1 T) to 30 μm (60 min, 0.4 T). Figure 30a shows a scanning electron microscopy (SEM) image of SiNWs grown at 0.1 T for 10 min. It consists of low density of SiNWs around 1 μm in length. High density of SiNWs with an average diameter in the range of 20–150 nm and 30 μm in length were obtained at 0.4 T for 60 min, leading to a nonuniform structured surface (Fig. 30b) Table 1.

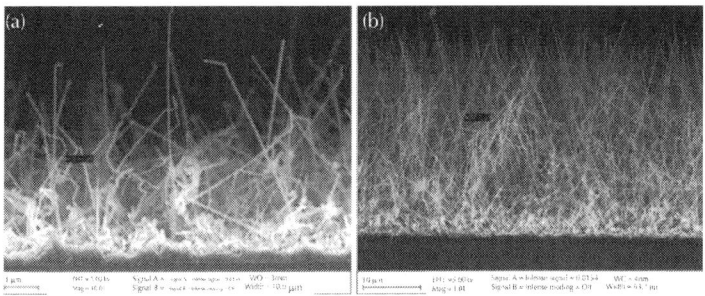

Figure 30: SEM images of silicon nanowires grown on a silicon wafer coated with a thin gold layer (4 nm) at 500 °C (a) P = 0.1 T, (b) P = 0.4 T. The silane flow is of 40 sccm, the time of growth is 60 min.

Table 1: Growth conditions of silicon nanowires (Q = 40 sccm, T = 500 °C)

Growth conditions of silicon nanowires (Q = 40 sccm, T = 500 °C)			
No.	Time (min)	Pressure (T)	Length (µm)
1	10	0.1	1
2	10	0.4	1
3	20	0.1	2.5
4	20	0.4	15
5	40	0.1	8
6	40	0.4	35
7	60	0.1	7
8	60	0.4	30

Verplanck et al.

Verplanck et al. Nanoscale Research Letters 2007 2:577-596, doi: 10.1007/s11671-007-9102-4

To achieve surface superhydrophobicity, the SiNWs were coated with a fluoropolymer C_4F_8 (60 nm thick), deposited using a plasma technique. All the resulting surfaces displayed liquid contact angle * around 164° for a saline solution (100 mM KCl) in oil (undecane) with almost no hysteresis, confirming that the droplet is in a Cassie state. Electrowetting in oil was performed on all surfaces, but a reversible behavior was only observed for the surface prepared using the process 8. When a voltage of 150 V_{rms} was applied, the apparent contact angle decreased down to 106° for a saline solution (100 mM KCl). When the tension was cut off, the effect is completely reversible. The drop returns to its initial position. Applied voltage leads to nonreversible wetting on the other surfaces (droplet trapped in a Wenzel configuration).

Same experiments have been carried out in air, on all the surfaces. Only the surface prepared using the process 8 allows a reversible electrowetting with electrowetting induced a maximum reversible decrease of the contact angle of 23° to reach 137° (starting from 160°). Turning off the voltage leads to a complete relaxation of the droplet (Fig. 31). This effect is ascribed to the high heterogeneity of the surface and trapped air under the droplet preventing to reach the Wenzel configuration [92].

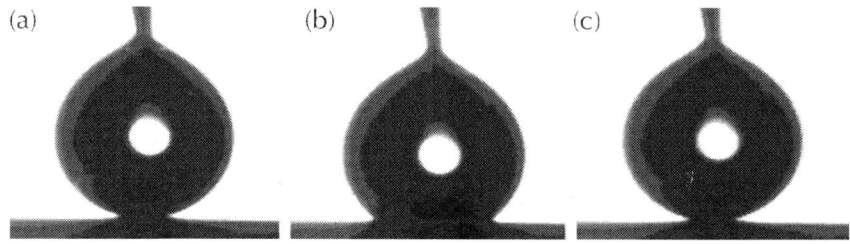

Figure 31: Reversible EWOD observed on a drop deposited on a superhydrophobic silicon nanowires surface. (a) No tension applied, (b) a 150 V_{rms} tension applied (f = 1 kHz), (c) the tension is cut, the drop returns to its initial state.

We have shown for the first time that reversible electrowetting is possible on superhydrophobic surfaces that display specific geometrical criteria as predicted by Bico [24]. Due to low hysteresis of the surface, we assume that small voltages could be sufficient for droplet displacement. We have previously demonstrated the possibility to use such surfaces as EWOD ground electrodes with hydrophobic electrodes for matrix-free mass-spectrometry analysis (DIOS analysis) [91]. The main advantages associated are a simple realization of hydrophilic and functionalized pads in the superhydrophobic surface, allowing analytes trapping with an enhancement of the liquid/surface interaction, and a subsequent analysis by matrix-free desorption/ionization MS-DIOS on these pads.

Integration of the superhydrophobic electrodes inside a microfluidic microsystem, allowing low voltage actuation of a biological analyte and DIOS analysis is currently under investigation in our laboratory. Furthermore, the utilization of textured surfaces could prevent from nonspecific sticking of bio particles, leading to an easy and efficient removal operation as compared to planar surface. Application such as particle sampling, concentration and analysis on superhydrophobic surfaces should be dedicated to environment control.

CONCLUSIONS

Among all the superhydrophobic surfaces displaying high roughness combined with low surface energy coating, trapping of air between the substrate and the liquid droplets is necessary to obtain a rolling ball effect (i.e. a quasi-null hysteresis). Associated to an effective way to switch the wettability properties of the surface, control of droplet displacement on superhydrophobic surface seems to be possible. Unfortunately, only few techniques based on optical, electrical, mechanical or magnetic phenomenon, lead to a reversible modification of surface wettability. Among these techniques, electrowetting on classical surfaces (i.e. hydrophobic) seems to be the more mature technology. This is particularly emphasized by recent results on EWOD droplet liquid pixel and by the very last improvement concerning optical lenses integrated inside commercialized cellular phones (varioptic.com). Combining the amazing properties of superhydrophobic surfaces with reliable EWOD devices will open new opportunities for designing systems with potential applications based on specific properties of theses surfaces, in particular in the field of lab-on-chip (preparation of highly functional microfluidic devices), optical devices and controlled self-cleaning surfaces. Concerning lab-on-chip devices, the most important effect expected, due to the quasi null hysteresis of these surfaces, is the liquid manipulation at very low tension voltage.

REFERENCES

1. L. Feng, S. Li, H. Li, J. Zhai, Y. Song, L. Jiang, D. Zhu, Angew. Chem. Int. Ed. 41, 1221 (2002)
2. L. Feng, Y. Song, J. Zhai, B. Liu, J. Xu, L. Jiang, D. Zhu, Angew. Chem. Int. Ed. 42, 800 (2003)
3. M. Cao, X. Song, J. Zhai, J. Wang, Y. Wang, J. Phys. Chem. B 110, 13072 (2006)
4. Y.C. Hong, H.S. Uhm, Appl. Phys. Lett. 88, 244101 (2006)
5. K. Tadanaga, N. Katata, T. Minami, J. Am. Ceram. Soc. 80, 1040 (1997)
6. E. Balaur, J.M. Macak, H. Tsuchiya, P. Schmuki, J. Mater. Chem. 15, 4488 (2005)
7. W. Chen, A.Y. Fadeev, M.C. Hsieh, D. Oner, J. Youngblood, T.J. McCarthy, Langmuir 15, 3395 (1999)
8. S.R. Coulson, I. Woodward, J.P.S. Badyal, S.A. Brewer, C.J. Willis, J. Phys. Chem. B 104, 8836 (2000)
9. R. Fürstner, W. Barthlott, Langmuir 21, 956 (2005)
10. J.-Y. Shiu, C.W. Kuo, P. Chen, C.Y. Mou, Chem. Mater. 16, 561 (2004)
11. T.J. McCarthy, D. Oner, Langmuir 16, 7777 (2000)
12. T. Young, Philos. Trans. R. Soc. Lond. 95, 65 (1805)
13. J. Fox, W. Zisman, J. Colloid Interface Sci. 5, 514 (1950)
14. W. Zisman, Chem. Ser. 43, 381 (1964)
15. P-G de Gennes, F. Brochard-Wyart, D. Quere. Gouttes, bulles, perles et ondes. (Belin, collection Echelles, Paris, 2002)
16. http://www.bath.ac.uk/ceos/insects3.html
17. R.N. Wenzel, Ind. Eng. Chem. 28, 988 (1936)

18. R.N. Wenzel, J. Phys. Colloid Chem. 53, 1466 (1949)
19. A.B.D. Cassie, S. Baxter, Trans. Faraday Soc. 40, 546 (1944)
20. R.E. Johnson, R.H. Dettre, Adv. Chem. Ser. 43, 112 (1964)
21. J. Bico, The`se, Université Paris VI (2000)
22. M. Nosonovsky, Langmuir 23, 3157 (2007)
23. C. Yang, U. Tartaglino, B.N.J. Persson, Phys. Rev. Lett. 97, 116103 (2006)
24. M. Callies, D. Quéré, Soft Matter 1, 55 (2005)
25. D. Quéré, Physique statistique, Images de la Physique, CNRS 239 (2005)
26. J. Bico, U. Thiele, D. Quéré, Colloids Surf. A 206, 41 (2002)
27. S. Shibuichi, T. Onda, N. Satoh, K. Tsujii, J. Phys. Chem. 100, 19512 (1996)
28. J. Bico, C. Marzolin, D. Quéré, Europhys. Lett. 47, 220 (1999)
29. Z. Yoshimitzu, A. Nakajima, T. Watanabe, K. Hashimoto, Langmuir 18, 5818 (2002)
30. A. Lafuma, D. Quéré, Nature Mater. 2, 457 (2003)
31. D. Quéré, A. Lafuma, J. Bico, Nanotechnology 14, 1109 (2003)
32. S. Herminghaus, Europhys. Lett. 52, 165 (2000)
33. N.A. Patankar, Langmuir 20, 8213 (2004)
34. L. Gao, T.J. McCarthy, Langmuir. 22, 2966 (2006)
35. K.K.S. Lau, J. Bico, K.B.K. Teo, M. Chlowalla, G.A.J. Amaratunga, W.I. Milne, G.H. McKinley, K.K. Gleason, Nano Lett. 3, 1701 (2003)
36. L. Feng, S. Li, Y. Li, H. Li, L. Zang, J. Zhai, Y. Song, B. Liu, L. Jiang, D. Zhu, Adv. Mater. 14, 1857 (2002)
37. H. Liu, J. Zhai, L. Jiang, Soft Matter. 2, 811 (2006)
38. J. Lee, B. He, N. Patankar, J. Micromech. Microeng. 15, 591 (2005)
39. T.H. Chen, Y.J. Chuang, C.C. Chieng, F.G. Tseng, J. Micromech. Microeng. 17, 489 (2007)

40. J. Zhang, X. Lu, W. Huang, Y. Han, Macromol. Rapid Commun. 26, 477 (2005)
41. M. Jin, X. Feng, L. Feng, T. Sun, J. Zhai, T. Li, L. Jiang, Adv. Mater. 17, 1977 (2005)
42. X. Hong, X. Gao, L. Jiang, J. Am. Chem. Soc. 129, 1478 (2007)
43. F. Zhou, W.T.S. Huck, Chem. Commun. 5999 (2005)
44. T.L. Sun, G.J. Wang, L. Feng et al., Angew. Chem. Int. Ed. 43, 357 (2004)
45. Q. Fu, G.V.R. Rao, S.B. Basame, D.J. Keller, K. Artyushkova, J.E. Fulghum, G.P. Lopez, J. Am. Chem. Soc. 126, 8904 (2004)
46. F. Xia, L. Feng, S. Wang, T. Sun, W. Song, W. Jiang, L. Jiang, Adv. Matter. 18, 432 (2006)
47. A. Athanassiou, M.I. Lygeraki, D. Pisignano, K. Lakiotaki, M. Varda, E. Mele, C. Fotakis, R. Cingolani, S.H. Anastasiadis, Langmuir 22, 2329 (2006)
48. H.S. Lim, J.T. Han, D. Kwak, M. Jin, K. Cho, J. Am. Chem. Soc. 128, 14458 (2006)
49. W. Zhu, X. Feng, L. Feng, L. Jiang, Chem. Commun. 2753 (2007)
50. X. Feng, L. Feng, M. Jin, J. Zhai, L. Jiang, D. Zhu, J. Am. Chem. Soc. 126, 62 (2004)
51. C. Takei, M. Nonogi, A. Hibara, T. Kitamori, H.B. Kim, Lab on Chip 7, 596 (2007)
52. E. Balaur, J.M. Macak, L. Taveira, P. Schmuki, Electrochem. Commun. 7, 1066 (2005)
53. Y. Coffinier, S. Janel, A. Addad, R. Blossey, L. Gengembre, E. Payen, R. Boukherroub, Langmuir 23, 1608 (2007)
54. G. Lippmann, Ann. Chim. Phys. 5, 494 (1875)
55. B. Berge, C. R. Acad. Sci. Paris série II 317, 157 (1993)
56. Y. Fouillet, D. Jary, A.G. Brachet, J. Berthier, R. Blervaque, L. Davoux, J.M. Roux, J.L. Achard, C. Peponnet, 4th International Conference on Nanochannels, Microchannels and Minichannels, Liemrick, Ireland, June 19–21 (2006)

57. A. Froumkine, Actualite´s Sci. Ind. 373, 1 (1936)
58. T.B. Jones, Langmuir 18, 4437 (2002)
59. K.H. Kang, Langmuir 18, 10318 (2002)
60. J. Buehrle, S. Herminghaus, F. Mugele, Phys. Rev. Lett. 91, 086101 (2003)
61. M. Bienia, M. Vallade, C. Quilliet, F. Mugele, Europhys. Lett. 74, 103 (2006)
62. F. Mugele, J. Buerhle, J. Phys.: Condens. Matter 19, 375112 (2007)
63. S. Kuiper, 5th International Meeting on Electrowetting, Rochester (NY, USA), May 31–June 2 (2006)
64. A. Torkelli, Droplet microfluidics on planar surface, ISBN 951- 38-6237-2 (2003)
65. H.J.J. Verheijen, M.W.J. Prins, Langmuir 15, 6616 (1999)
66. M. Vallet, M. Vallade, B. Berge, Eur. Phys. J. B 11, 583 (1999)
67. B. Shapiro, H. Moon, R.L. Garrell, C.J. Kim, J. Appl. Phys. 93, 5794 (2003)
68. F. Mugele, J.C. Baret, J. Phys.: Condens. Matter. 17, R705 (2005)
69. R.D. Fair, Microfluid Nanofluid 3, 245 (2007)
70. B. Berge, J. Peseux, Eur. Phys. J. E 3, 159 (2000)
71. Varioptic, http://www.varioptic.com
72. H. Pellat, C. R. Acad. Sci. Paris. 119, 691 (1895)
73. N.R. Smith, D.C. Abeysinghe, J.W. Haus, J. Heikenfeld, Optics Express. 14, 6557 (2006)
74. Liquavista, http://www.liquavista.com
75. R.A. Hayes, B.J. Feenstra, Nature 425, 383 (2003)
76. J. Heikenfeld, A.J. Steckl, Appl. Phys. Lett. 86, 011105 (2005)
77. S.K. Cho, H. Moon, C.J. Kim, J. Microelec. Sys. 12, 70 (2003)
78. J. Berthier, P. Silberzan, Microfluidics for Biotechnology (Artech House Publishers 2005)
79. M.G. Pollack, R.B. Fair, A.D. Shenderov, Appl. Phys. Lett. 77, 1725 (2000)

80. M.G. Polack, A.D. Shenderov, R.B. Fair, Lab Chip 2, 101 (2002)
81. V. Srinivasan, V.K. Pamula, R.B. Fair, Lab Chip. 4, 310 (2004)
82. A.R. Wheeler, H. Moon, C.A. Bird, R.R. Ogorzalek Loo, C.J. Kim, J.A. Loo, R.L. Garrell, Anal. Chem. 77, 534 (2005)
83. F. Caron, J.-C. Fourrier, C. Druon, P. Tabourier, French Patent N FR 0406080 issued on 2005 Nanoscale Res Lett (2007) 2:577–596 595 123
84. Y. Fouillet, H. Jeanson, I. Chartier, A. Buguin, P. Silberzan, Houille blanche, Revue Internationale de l'Eau 4, 37 (2003)
85. T.N. Krupenkin, J.A. Taylor, T.M. Schneider, S. Yang, Langmuir 20, 3824 (2004)
86. T. Krupenkin, J.A. Taylor, P. Kolodner, M. Hodes, Bell Labs Tech. J. 10, 161 (2005)
87. D.L. Herbertson, C.R. Evans, N.J. Shirtcliffe, G. McHale, M.I. Newton, Sens. Actuators A 130, 189 (2006)
88. M.S. Dhindsa, N.R. Smith, J. Heikenfeld, P.D. Rack, J.D. Fowlkes, M.J. Doktycz, A.V. Melechko, M.L. Simpson, Langmuir 22, 9030 (2006)
89. A. Klingner, F. Mugele, J. Appl. Phys. 95, 2918 (2004)
90. K. Bhat, J. Heikenfeld, M. Agarwal, Y. Lvov, K. Varahramyan, Appl. Phys. Lett. 91, 024103 (2007)
91. N. Verplanck, Y. Coffinier, M. Wisztorski, G. Piret, C. Delhaye, V. Thomy, I. Fournier, J.-C. Camart., P. Tabourier, R. Boukherroub, The 10th International Conference on Miniaturized Systems for Chemistry and Life Sciences (ITAS Tokyo) 771 (2006)
92. N. Verplanck, Y. Coffinier, E. Galopin, J.-C. Camart, V. Thomy, R. Boukherroub, Nano Lett. 3, 813 (2007)

Chapter 5

ELIXYS - A Fully Automated, Three-reactor High-pressure Radiosynthesizer for Development and Routine Production of Diverse PET Tracers

Mark Lazari[1,2,3], Kevin M Quinn[2,3], Shane B Claggett[2,3], Jeffrey Collins[2,3], Gaurav J Shah[2,3,4], Henry E Herman[2,3], Brandon Maraglia[2,3,4], Michael E Phelps[2,3], Melissa D Moore[2,3,4], and R Michael van Dam[1,2,3]

[1]Department of Bioengineering, Henry Samueli School of Engineering, UCLA, Los Angeles, CA 90095, USA

[2]Department of Molecular and Medical Pharmacology, David Geffen School of Medicine, UCLA, 4323 CNSI, 570 Westwood

Plaza, Building 114, Los Angeles, CA 90095, USA

[3]Crump Institute for Molecular Imaging, David Geffen School of Medicine, UCLA, Los Angeles, CA 90095, USA

[4]Sofie Biosciences, Inc., Culver City, CA 90230, USA

ABSTRACT

Background

Automated radiosynthesizers are vital for routine production of positron-emission tomography tracers to minimize radiation exposure to operators and to ensure reproducible synthesis yields. The recent trend in the synthesizer industry towards the use of disposable kits aims to simplify setup and operation for the user, but often introduces several limitations related to temperature and chemical compatibility, thus requiring reoptimization of protocols developed on non-cassette-based systems. Radiochemists would benefit from a single hybrid system that provides tremendous flexibility for development and optimization of reaction conditions while also providing a pathway to simple, cassette-based production of diverse tracers.

Methods

We have designed, built, and tested an automated three-reactor radiosynthesizer (ELIXYS) to provide a flexible radiosynthesis platform suitable for both tracer development and routine production. The synthesizer is capable of performing high-pressure and high-temperature reactions by eliminating permanent tubing and valve connections to the reaction vessel. Each of the three movable reactors can seal against different locations on disposable cassettes to carry out different functions such as sealed reactions, evaporations, and reagent addition. A reagent and gas handling robot moves sealed reagent vials from storage locations in the cassette to addition positions and also dynamically provides vacuum and inert gas to ports on the cassette. The software integrates these automated features into chemistry unit operations (e.g., React, Evaporate, Add) to intuitively create synthesis protocols. 2-Deoxy-2-[^{18}F] fluoro-5-methyl-β-L-arabinofuranosyluracil (L-[^{18}F] FMAU) and 2-deoxy-2-[^{18}F] fluoro-β-D-arabinofuranosylcytosine (D-[^{18}F] FAC) were synthesized to validate the system.

Results

L-[^{18}F]FMAU and D-[^{18}F]FAC were successfully synthesized in 165 and 170 min, respectively, with decay-corrected radiochemical yields of 46% ± 1% (n = 6) and 31% ± 5% (n = 6), respectively. The yield, repeatability, and synthesis time are comparable to, or better than, other reports. D-[^{18}F] FAC produced by ELIXYS and another manually operated apparatus exhibited similar biodistribution in wild-type mice.

Conclusion

The ELIXYS automated radiosynthesizer is capable of performing radiosyntheses requiring demanding conditions: up to three reaction vessels, high temperatures, high pressures, and sensitive reagents. Such flexibility facilitates tracer development and the ability to synthesize multiple tracers on the same system without customization or replumbing. The disposable cassette approach simplifies the transition from development to production.

BACKGROUND

Positron-emission tomography (PET) has opened the door to in vivo imaging for the purposes of non-invasive disease detection, cancer staging, and drug efficacy screening [1]. 2-[^{18}F] fluoro-2-deoxy-D-glucose ([^{18}F] FDG) is the most commonly utilized PET tracer due to its relative ease of production, manageable half-life, and ubiquitous application [2, 3]. The increased demand for [^{18}F]FDG has led to the development of many automated radiosynthesizers to lower its cost, enable its production at many different sites, and reduce the radiation exposure to the radiochemist [4,5].

Though automated synthesis of [^{18}F] FDG is extremely valuable, there are many ^{18}F-labeled PET tracers that await an automated synthesizer to streamline their production [6]. Some of these tracers require high pressures, complicated chemistry, and/or corrosive reagents that make automation difficult. For example, nucleoside analogs that have been used in imaging cell proliferation and reporter gene expression [7-9] and as possible screening agents for chemotherapy drug efficacy [10] often require high-temperature reactions in volatile solvents. Several attempts have been made to automate the syntheses of these tracers on commercially available radiosynthesizers. Often, these attempts have required modifications to the chemistry (e.g., use of alternative solvents or reduced temperatures) to reduce the pressures involved and avoid exceeding the limitations of the radiosynthesizers [8, 11-16].

To overcome these synthesizer limitations, we previously developed a platform with movable components that seals the reaction vessel against an inert stopper during reactions to avoid exposure of tubing and valves to high pressures [17]. We further developed this system into a modular computer-controlled platform [18] and demonstrated the successful synthesis of 2-deoxy-2-[^{18}F] fluoro-5-methyl- -L-arabinofuranosyluracil (L-[^{18}F] FMAU) and 2-deoxy-2-[^{18}F] fluoro- -D-arabinofuranosylcytosine (D-[^{18}F] FAC). We describe here the integration of three of these modules into an automated synthesizer [19] and the addition of numerous improvements and features to increase the reliability and user-friendliness of the system. The ELIXYS three-reactor synthesizer is designed to use disposable cassettes for ease of setup and operation and to facilitate rapid transition from tracer development to routine production. The system also has an integrated, automated reagent handling robot to deliver sensitive reagents from sealed vials on demand. This paper describes in detail this radiosynthesizer and its characterization and validation via the synthesis of D-[^{18}F] FAC and

L-[^{18}F] FMAU. The details of the software interface were published as a companion article [20].

METHODS

Apparatus

The hardware and software of the ELIXYS radiosynthesizer were designed to be user-friendly and accommodate a wide variety of synthesis protocols by allowing the user to customize the reagents, reaction times, temperatures, and intermediate purifications in a manner that focuses on organizing automated synthesis steps into 'unit operations.' These chemistry operations serve as intuitive building blocks from which diverse syntheses can be constructed without the need for reconfiguring the instrument to perform different syntheses.

The radiosynthesizer (Figure 1) has three key components working in concert: a set of three reactors (Figure 2), a reagent and gas handling robot (Figure 3), and disposable cassettes (Figure 4). The cassettes store reagents in sealed vials, act as the primary fluid path for both reagents and gas flow, and have a rubber gasket affixed to the bottom for sealing the top of glass reaction vessels. Cassettes accelerate setup, eliminate the need for cleaning, and facilitate a natural transition from tracer development to routine production. The reactor subassemblies provide temperature control of the reaction vessel and movement of the vessel to various positions beneath the cassette designed for evaporations, sealed reactions, reagent addition, and transfer of product. Once aligned at the proper position, the reactor is raised to seal the top of the vessel against the gasket on the underside of the cassette. The two-axis reagent and gas handling robot supplies inert gas (to drive fluid movement and assist with evaporations) and vacuum (to remove vapor during evaporations) to special interfaces on the top of the cassettes. A gripper mounted to the same robot manipulates reagent vials between storage and addition positions.

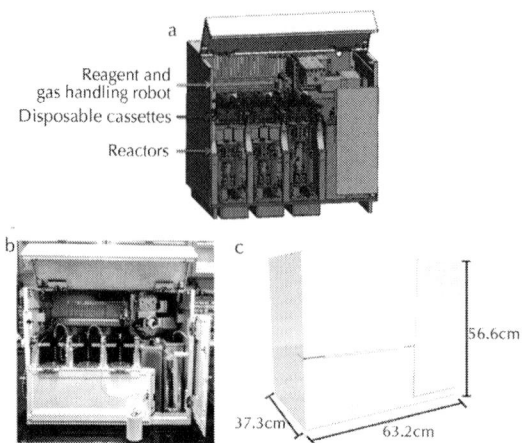

Figure 1: ELIXYS synthesis module. (a) Schematic, (b) photograph, and (c) enclosure with dimensions of the ELIXYS radiosynthesizer. The three reactors can be moved back and forth and raised to seal the vessel against various positions on the gasket at the bottom of the disposable cassettes to perform various unit operations. A reagent and gas handling robot manipulates reagent vials and supplies inert gas and vacuum for all three cassettes.

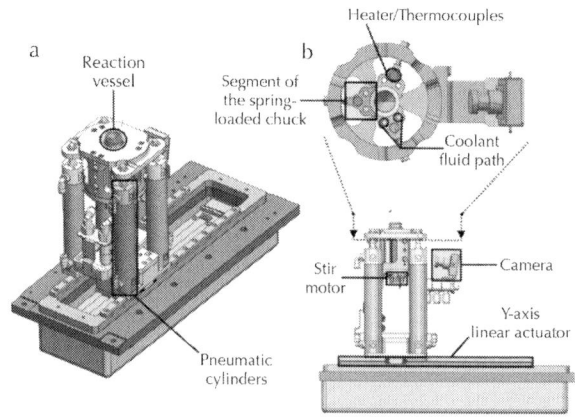

Figure 2: Detailed view of a single reactor. (a) 3D view of the reactor sub-assembly. (b) Side view (bottom) with a cross section through the spring-loaded chuck (top).

ELIXYS - A Fully Automated, Three-reactor High-pressure... 141

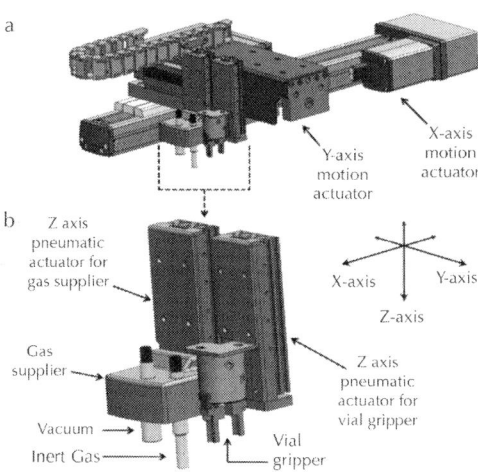

Figure 3: The reagent and gas handling robot. (a) Overview. (b)Close-up of the vial gripper and gas supplier head. The head moves in the x- and y-directions to access reagent vial positions and inert gas and vacuum ports on the three cassettes. The vial gripper and gas supplier can independently move in the z-direction.

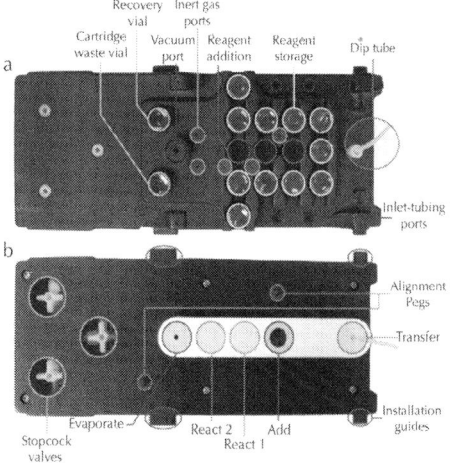

Figure 4: Overview of the disposable cassette. (a) The top view of the disposable cassette illustrates the locations for reagent addition and storage along with gas ports for the gas supplier to provide vacuum and inert gas. Tubing extends out of the inlet-tubing ports for radioisotope and external

additions. (b) Bottom view. The reactor seals against various positions on a PTFE-coated silicone gasket underneath the cassette to perform various unit operations (Evaporate, React, Add, and Transfer). Each cassette contains three stopcock valves and has a dip tube for transferring products to purification cartridges, subsequent cassettes, or the HPLC injection valve.

Reactor

A 5-mL glass V-vial (W986259NG, Wheaton, and Millville, NJ, USA) is placed into the reactor and held within a three-segment spring-loaded 'chuck' (Figure 2). The segments press firmly against the vial to ensure excellent thermal contact and thus efficient heat exchange between the reactor and the glass vial. Each segment has one 100-W cartridge heater (CIR-1021-120V-100W-ST-A, Valin, and San Jose, CA, USA) and a K-type thermocouple (HTTC72-K-116U-1.25-UNGR, Omega Engineering; Stamford, CT, USA) for individual feedback control of the reactor temperature. The maximum reactor temperature is 185°C, but we are investigating alternative types of fittings to enable an even higher operating temperature. Since we typically observed very similar temperature response in all three segments, the reactor temperature at any given moment is considered equal to the average of the three temperature readings from the thermocouples. Active liquid cooling is achieved by pumping room-temperature coolant (propylene/ethylene glycol and water mixture) through cooling channels in all three reactors in series using a liquid pump (8030-863-236, Steam Brite, San Antonio, TX, USA) and then through a radiator with three 140-mm fans (HX-CU1403V, Frozen CPU, East Rochester, NY, USA).

Each reactor is situated beneath a cassette and actuated back and forth among various positions (Figures 4 and 5). In each position, the reactor can be raised using pneumatic cylinders to seal the top of the reaction vessel against a portion of the gasket affixed to the bottom of the cassette. This movement allows for the reaction vessel and fluid path to be dynamically configured for different unit operations (Figure 5). For example, in one position, tubing is present to deliver reagents to the reactor; in another position, there is no tubing, allowing for a reaction under sealed conditions. Permanent tubing and valve connections to the reaction vessel are the root cause of the reaction pressure limitations of most synthesizers [17]. The ability to move the reaction vessel to a dedicated sealed reaction position eliminates these limitations and enables compatibility with higher pressures. To ensure reliable operation, the position of the reaction vessel is monitored via feedback from the linear actuator and the raised or lowered state is detected with Hall Effect sensors (D-M9NWL, SMC Corporation, Noblesville, IN, USA).

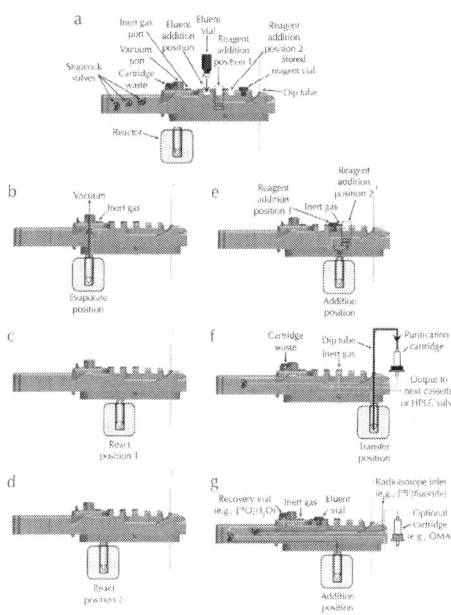

Figure 5: Fluid diagram for each unit operation. (a) Side profile schematic of the disposable cassette. The remaining figures show the schematic of cassette fluid paths for each unit operation. (b) Evaporate. Gas supplier provides vacuum and inert gas flow while the reactor is heated. (c) React1. First fully sealed reaction position. (d) React2. Second fully sealed reaction position. (e) Add. Vial gripper presses a reagent vial into one of two addition positions where two needles pierce the vial's septum; one needle allows inert gas flow from the gas supplier through the inert gas port, and the other needle allows the reagent to flow into the reaction vessel. (f) Transfer. Contents of one reaction vessel can be transferred to

another cassette, the HPLC valve, or to a purification cartridge. (g) Radio-isotope handling. [^{18}F] fluoride trap and release can be done using two of the built in stopcock valves.

We have mounted a camera (PC213XS, Super Circuits, and Austin, TX, USA) behind the reactor, which is especially helpful during synthesis development to monitor liquid levels during evaporations, to observe visual cues for reaction progression, to confirm reagent additions and transfers, and for visual inspection of the eluate post purification. The final component of the reactor is a magnet mounted on a DC motor (803-313-5858, KALEJA Elektronik GmbH, Alfdorf, Germany) to rotate a removable magnetic stir bar inside the reaction vessel.

Reagent and Gas Handling Robot

A pneumatically actuated vial gripper (Figure 3) that is affixed to a three-axis motion system grasps reagent vials and moves them from storage locations to addition locations (and vice versa) in the cassettes. The x- and y-axis motions are performed by a pair of linear actuators. The z-axis motion is accomplished with a pneumatic cylinder. Sealed reagent vials are installed upside-down in the cassettes (Figure 5).

A gas supplier, comprised of inert gas and vacuum supply ports, is mounted on a second pneumatic z-axis actuator affixed near the vial gripper (Figure 3). In the previous prototype to the ELIXYS [21], the cassettes each had multiple inputs on their bottom surface for inert gas and vacuum that mated with corresponding ports on the synthesizer. This system was prone to failure due to the large number of connections and the difficulty in maintaining reliable gas-tight seals during operation. With a single, movable gas supplying robot in ELIXYS, a large number of valves and seals are eliminated, increasing the reliability of the system. The gas supplier is lowered to engage respective vacuum and inert gas ports on top of the cassette (Figure 4). The cassettes contain a rubber gasket around each port to form a gas-tight seal with the port on the robot. The vacuum port is mounted at a height above the inert gas port and on a spring-loaded mechanism to ensure proper sealing of both ports when needed and to avoid collision of the vacuum portion of the gas supplier when vacuum is not needed.

The use of Hall Effect sensors as feedback mechanisms on z-axis pneumatic actuators and the vial gripper enable detection of missing reagent vials and prevent x- and y-axis motions if the vial gripper and gas supplier are not in their raised, clearance positions. An in-line check valve (CI-5C, Bio Chem Fluidics, and Boonton, NJ, USA) is installed on the inert gas line close to the delivery point to eliminate backflow of vapor. A cold trap (CG451501, Chemglass, Vineland, NJ, USA), cooled in a small Dewar (10-195A, Fisher Scientific, Pittsburg, PA, USA), typically with a mixture of dry ice and methanol, is installed in-line between the vacuum port and the integrated vacuum pump (VP0140-V1006-D2-0511, Medo USA Inc., Roselle, IL, USA) with a digital vacuum gauge (ZSE30-N7L, SMC Corporation).

Disposable Cassettes

Cassettes (Figure 4) are designed to contain all disposable components and fluid paths, eliminating the need to clean or customize the apparatus between syntheses. These molded polyurethane cassettes contain stainless steel needles (Vita Needle, Needham, MA, USA), tubing, chemically inert three-way stopcock valves (EW-31200-80, Cole-Parmer, Vernon Hills, IL, USA), and a custom polytetrafluoroethylene (PTFE)-coated silicone gasket (Specialty Silicone Products, Inc., Ballston Spa, NY, USA, and Cannon Gasket, Upland, CA, USA) against which the reaction vessel is sealed. The fluid paths are shown in Figure 5. Preassembled cassettes slide along rails into the ELIXYS system which are then locked down by clamps. Accurate positioning is ensured by the mating of alignment features and by the engagement of stopcock valves with adapters affixed to rotary pneumatic actuators (CRB2BW20-180S, SMC Corporation). Each cassette has 11 reagent vial storage positions that each house one 13-mm crimped septum-cap vial (with a maximum volume of 3 mL) in an inverted configuration. Three additional vial positions contain dual upward-pointing needles to pierce the septa for delivery of fluids from the reagent vials. One needle in each position is used for fluid delivery; the other connects to an inert gas port on top of the cassette, which allows pressurization of the vial by the gas supplier. In the two reagent addition positions, the fluid delivery needles output directly to the underside of the cassette where the reaction vessel is sealed for reagent addition. The fluid delivery needle in the third position (for eluent addition) is connected via internal tubing to a stopcock valve.

Two of the stopcock valves are used to reconfigure the internal fluid paths to perform cartridge trap and release, such as for the preparation of an incoming radioisotope, e.g., solvent exchange of [^{18}F] fluoride to recover [^{18}O] H_2O. The upstream valve selects between incoming fluid from an external addition line (trapping) or the eluent addition position (release). The output of the first stopcock is connected via tubing to a cartridge. The output of the cartridge is connected to a second tube that feeds into a second stopcock valve that selects whether the fluid is directed to a built-in collection vial (trapping) or a line into the position where the reaction vessel is sealed for reagent addition (release). The third stopcock valve is used for cartridge purification of the crude product before transfer to the next reaction vessel. The purification cartridge is installed between the dip tube (for removal of crude product from the reaction vessel) and the tube leading to the stopcock. The outputs of the stopcock are connected to a built-in waste vial (trapping, washing) or an external output line (release) that can be plugged into the next cassette or the high-performance liquid chromatography (HPLC) injection valve. Cartridges can be mounted on clips near the front of the cassettes for convenience.

Control System

Supporting electronics, pneumatics, and cooling system are enclosed in a separate control system (Figure 6). The system is controlled by a Linux server, which communicates with a programmable logic controller (PLC; CJ2M-CPU31, Omron, Kyoto, Japan) over Ethernet, which in turn drives most of the subsystems including linear actuators, pneumatics, cooling, heating, stirring, and HPLC injection, in addition to reading radioactivity detectors affixed to each reactor. The PLC accomplishes this through several expansion modules (CJ1W-DRM21, CJ1W-AD081-V1, CJ1W-ID261, CJ1W-DA08V, CJ1W-OD261, CJ1W-TC001, and Omron).

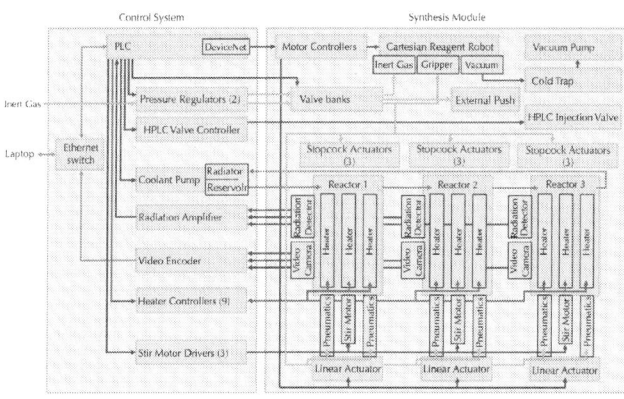

Figure 6: ELIXYS control system schematic. Schematic of control components, actuators, and sensors in the control system and synthesis module.

Five motor controllers are connected to a RoboNET network controller gateway unit (RGW-DV, IAI America Inc., Torrance, CA, USA) in the synthesis module, which is in turn controlled by the PLC. Two of these are pulse motor controllers (RPCON-42P, IAI America Inc.) that drive the x- and y-axes of the reagent and gas handling robot, a 350- and 100-mm-stroke two-axis linear servomotor (RCP2-SS7R-I-42P-12-350-P1-007L-ML-SP, RCP3-TA7R-I-42P-6-100-P1-N-ML, IAI America Inc.). The other three controllers are linear servomotor controllers (RACON-5, IAI America Inc.) driving the linear servomotor (RCP3-SA3R-I-28P-4-200-P1-P-ML, IAI America Inc.) for the y-axis motion of each reactor.

A source of inert gas is dynamically regulated in software from >60 psig down to two different pressures by two analog pressure regulators (ITV1030-31N2L4-Q, SMC Corporation). One pressure line drives the pneumatic actuators (typically set to 60 psig), and the other pressure line drives gas flow for liquid transfers and evaporation (typically 3 to 15 psig). The two lines are distributed to actuators and the gas supplier through solenoid valve banks located in the synthesis module. The higher pressure line is used to raise and lower the reaction vessels using pneumatic cylinders (NCDGBN20-0300, SMC Corporation), turn the stopcock valves via the rotary pneumatic actuators (CRB2BW20-180S, SMC Corporation), raise and lower the two z-axis actuators (MXS8-50, SMC Corporation) for the vial gripper and gas supplier, and open and close the vial gripper (MHS2-16D, SMC Corporation). The lower pressure line feeds into the gas supplier that seals to the gas inlet gaskets on top of the cassettes, and an external line that can be used to transfer [^{18}F] fluoride from a source vial into the anion exchange cartridge on the cassette, for example.

The control system also houses a number of other components: the solid-state relays (G6B-4BNDDC12, Omron) to switch the heaters on and off for reactor temperature control, the cooling system (coolant pump, reservoir, and radiator fans), a video server (VS8401, Vivotek, San Jose, CA, USA) to encode the analog signals from the reactor cameras into video streams available to the Linux server via Ethernet, and an electronically controlled HPLC injection valve (MHP7900-500-1, Rheodyne, Rohnert Park, CA, USA) connected to a separate semi-preparative HPLC system. Loading of the HPLC loop is performed manually but in the future will be performed automatically [22].

System Operations

The ELIXYS performs automated syntheses by completing a sequence of chemistry unit operations (Table 1). The interactions among the ELIXYS subsystems and disposable cassettes to carry out each operation are described in the following subsections.

Table 1: List of unit operations available to build a synthesis sequence

Unit operation	Description of function
INITIALIZE	Initializes hardware
TRAPF18	Trap [^{18}F]fluoride from cyclotron or pre-loaded external vial
ELUTEF18	Elute [^{18}F]fluoride with a reagent from the cassette
ADD	Add a reagent from any cassette
EVAPORATE	Evaporate the contents of a reactor
REACT	Fully seal a reaction vessel for a reaction
TRANSFER	Transfer solvents and reaction products from one reaction vessel to another, often using purification cartridges in between
TRANSFERTOHPLC[a]	Transfers the contents of the reaction vessel to the HPLC injection loop
EXTERNALADD	Move a reactor to its add position for externally adding a reagent
MIX	Mix the contents of a reaction vessel
INSTALL	Move a reactor to a position for reaction vessel removal and/or installation
MEASURERADIATION[a]	Measures the radiation level of a given reactor

[a]These unit operations are currently under development.

Lazari et al.

Lazari et al. EJNMMI Research 2013 3:52   doi: 10.1186/2191-219X-3-52

Radioisotope Handling

For radiochemistry with [^{18}F] fluoride, a preconditioned quaternary methylammonium (QMA) cartridge is installed with Luer fittings between two tubes coming from the cassette, and the source of [^{18}F] fluoride (vial or cyclotron) is connected to another tube. (If a vial is used, it is pressurized by an inert gas line controlled by the system.) During trapping, the [^{18}F] fluoride source solution flows through the cartridge and [^{18}O] H$_2$O is collected in a recovery vial. To perform elution, the two stopcock valves are switched and the reagent and gas handling robot drives the eluent from the eluent addition position of the cassette through the cartridge and into the reaction vessel. Multiple elutions or rinses can be performed to increase the efficiency of [^{18}F] fluoride collection. PEEK tubing was used for all fluid paths involving [^{18}F] fluoride to maximize specific activity [23]. For other radioisotopes, a cartridge may not be necessary and can be bypassed. Radioisotopes may be added to any of the three reactors independently.

Reagent Handling

To add a particular reagent, the vial gripper moves to the reagent storage position and then lowers itself to a position where it can grasp and lift the vial before moving it to the designated reagent addition location on the specified cassette. To deliver the reagent to the reaction vessel, the gas supplier lowers, the inert gas valve opens, and the vial gripper lowers to place the vial down onto a pair of needles in one of the two reagent addition positions or in the eluent addition position, pressurizing the vial and transferring its contents. The required time for addition of a reagent is generally determined by repeatedly measuring the time needed for complete transfer of the desired liquid and volume at the desired pressure, taking the maximum time value, and multiplying by a safety factor. The entire contents of the reagent vial are delivered at once. After addition is complete, the vial gripper lifts the empty reagent vial, the gas supplier disengages, and the vial is returned to its original storage position. The low-level steps of the Add unit operation are summarized in: Figure 1.

Like all fluidic systems, there are losses associated with dead volumes during liquid transfers. Initial characterization revealed that a small amount (120 ± 20 µL, n = 120) of the liquid remains in the reagent vial after addition. To account for this loss, additional reagent can be loaded beforehand into the crimped vials.

Reactions

To maintain high internal pressure during superheated reactions, the reaction vessel is sealed by firmly pressing upward against the gasket on the bottom of the cassette. Each cassette has two independent reaction positions to support up to two separate sealed reactions in each reaction vessel (Figure 5). To characterize the seal integrity, approximately 1 mL of anhydrous acetonitrile was sealed and heated at 165°C for 1 h. In all experiments, <14 µL of volume was lost (<1.5%). We believe the actual loss of vapor to be even lower because often some small drops of condensed solvent were observed on the gasket where the reaction vessel had been sealed.

Using a hypodermic needle thermocouple (HN-7-K-TEF, J-KEM Scientific, and Saint Louis, MO, USA) pierced through the gasket, we have also compared the internal liquid temperature profile of the reaction vessel contents in the ELIXYS to the profile inside the same vessel immersed in a traditional preheated oil bath. We found the heating rates and internal liquid temperature to be comparable, but the active liquid cooling of the ELIXYS results in a more rapid decrease in temperature after heating.

After reagents are loaded into the reaction vessel, a reaction can be performed by sealing the vessel against a sealing position on the gasket of the cassette (Figures 4 and 5). The reactor is then heated to the desired temperature, with optional stirring. Once the desired elevated temperature is reached in the heating jacket, heating and stirring are continued for the desired reaction time. After this time elapses, the heaters are turned off and the cooling pump is activated until the heating jacket reaches the desired reduced temperature. Additional cooling of the reaction vessel is necessary to ensure the internal liquid temperature is sufficiently lowered; the desired additional time for cooling can be programmed in the software.

Evaporations

Evaporation of solvents occurs by sealing the reaction vessel against the gasket of the cassette at the evaporate position (Figures

4 and 5). The vessel is heated with the option of stirring, and the gas supplier provides both vacuum (to remove vapor) and inert gas (to assist with vapor removal) through the ports on the cassette. The required time for evaporation is generally determined by measuring the maximum time needed for complete evaporation of the solvent from the desired mixture and multiplying by a safety factor. After the desired evaporation time, the reactor is cooled. The low-level steps of the Evaporate unit operation are summarized.

Transfer and Purification

Sep-Pak purification cartridges (Sep-Pak), e.g., silica, C18, etc., are connected to designated Luer fittings on the cassette. A dip tube (e.g., made of Teflon® tubing with an outside diameter (OD) of 1/8 in.) is built into the cassette to act as the fluid path for the transfer of crude products. Tubing with an OD of 1/8 in. is necessary when transferring synthesis products that produce precipitates, but 1/16 in. tubing may be installed for smaller volume transfers. The transfer unit operation begins with the reaction vessel sealing against the transfer position on the cassette (Figures 4 and5). The gas supplier provides inert gas to pressurize the reaction vessel. This moves the fluid through the dip tube and to the Sep-Pak. After the Sep-Pak, a dedicated stopcock valve in the cassette switches between a fluid path towards a waste collection vial installed on the cassette and a tube that can be plumbed to the input of the next cassette. Often, the first step is to trap the crude product onto the Sep-Pak and allow the residual solution to collect in the waste container. The stopcock position is switched, and elution of the desired product into the next reaction vessel is then performed by adding the elution solvent to the first reaction vessel and repeating the transfer unit operation to elute the product from the Sep-Pak.

Radiosynthesis

Materials

No-carrier-added [^{18}F] fluoride was produced by the (p,n) reaction of [^{18}O]H$_2$O (98% isotopic purity, Medical Isotopes, Pelham, NH, USA) in a RDS-112 cyclotron (Siemens, Knoxville, TN, USA) at 11 MeV using a 1-mL tantalum target with Havar foil. 2-O-(Trifluoromethylsulfonyl)-1,3,5-tri-O-benzoyl-alpha-D-ribo-furanose, 2-O-(trifluoromethylsulfonyl)-1,3,5-tri-O-benzoyl-alpha-L-ribo-furanose, bis(trimethylsilyl)cytosine, and 5-methyl-2,4-bis[(trimethylsilyl)oxy]pyrimidine were purchased from ABX (Radeberg, Germany). Ethanol (200 proof) was purchased from the UCLA Chemistry Department (Los Angeles, CA, USA). Hydrochloric acid (1 N) was purchased from Fisher Scientific (Pittsburg, PA, USA). Anhydrous grade solvents and all other reagents were purchased from Sigma-Aldrich (Milwaukee, WI, USA). All reagents were used as received. QMA (WAT023525) and silica cartridges (WAT020520 and WAT043400) were purchased from Waters (Milford, MA, USA). The QMA cartridge was preconditioned with 10 mL of 1 M potassium bicarbonate followed by 10 mL of 0.1-µm-filtered 18 MΩ water, and the silica cartridges were preconditioned with 10 mL of anhydrous hexane.

Chromatography

Semi-preparative HPLC was performed with a WellChrom K-501 HPLC pump (Knauer, Berlin, Germany), reversed-phase Gemini-NX column (5 µm, 10 × 250 mm, Phenomenex, Torrance, CA, USA), UV detector (254 nm, WellChrom Spectro-Photometer K-2501, Knauer), and gamma radiation detector and counter (B-FC-3300 and B-FC-1000; Bioscan Inc., Washington, DC, USA). The mobile phase for D-[^{18}F]FAC was 1% ethanol in 10 mM ammonium phosphate monobasic (flow rate 5 mL/min; retention time 15 min) and for L-[^{18}F]FMAU was 4% acetonitrile in 50 mM ammonium acetate (flow rate 5 mL/min; retention time 20 min). Analytical HPLC was performed on a Knauer Smartline HPLC system with a Phenomenex reversed-phase Luna column (5 μm, 4.6 × 250 mm) with in-line Knauer UV (254 nm) and gamma radiation coincidence detector and counter (B-FC-4100 and B-FC-1000). The analytical HPLC mobile phase for D-[^{18}F]FAC was 10% ethanol in 50 mM ammonium acetate (flow rate 1 mL/min; retention time 4 min) and for L-[^{18}F]FMAU was 10% acetonitrile in 50 mM ammonium acetate (flow rate 1 mL/min; retention time 7 min). All chromatograms were collected using a GinaStar analog-to-digital converter (raytest USA, Inc., Wilmington, NC, USA) and GinaStar software (raytest USA, Inc.) running on a PC.

Synthesis Protocol

Synthesis protocols for D-[^{18}F]FAC and L-[^{18}F]FMAU (Figure 7) were nearly identical, differing only in precursors and HPLC mobile phases, and were programmed using the ELIXYS drag-and-drop software interface [20]. The protocols are a minor adaptation from the literature [17, 18]. A summary of the reagents and unit operations used to synthesize the tracers can be found in 1 Upon completion of each synthesis, the crude product was purified by semi-preparative HPLC and the desired product (beta form, structures c and f in Figure 7) was collected and a sample taken for verification and specific activity analysis by analytical HPLC.

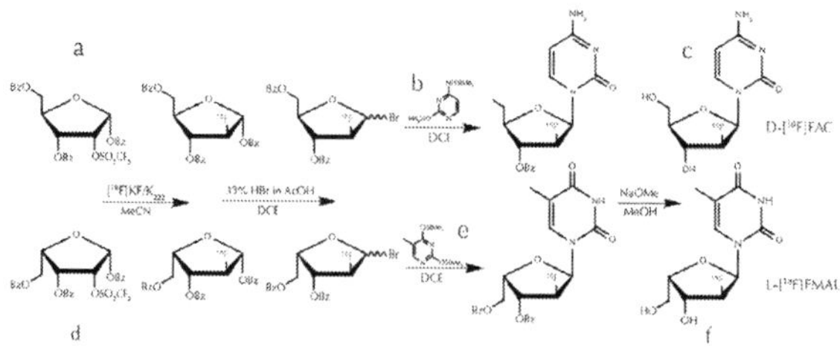

Figure 7: Reaction mechanisms of D-[18 F] FAC and L-[18F] FMAU. Syntheses of D-[^{18}F] FAC (top) and L-[^{18}F] FMAU (bottom). Synthesis protocol of the two tracers differs only in the ribose sugar (a vs. d) and base cou-

pling (b vs. e) precursors. During HPLC purification, only the beta form (shown here) are collected as final products (c and f).

In Vivo Imaging

Four conscious C57BL/6 mice were injected at the UCLA Ahmanson Translational Imaging Division with 0.74 MBq (20 µCi) of D-[^{18}F] FAC (tail vein, 60-min uptake) produced on the ELIXYS. Two days afterward, the same four mice were injected with D-[^{18}F] FAC produced at the UCLA Biomedical Cyclotron Facility using a manually operated apparatus [17]. Before scanning, mice were anesthetized with 2% isoflurane and placed in a dedicated imaging chamber designed for use with both the PET and CT systems. Ten-minute whole-body PET images were acquired using a GENISYS[4] (Sofie Biosciences Inc., Culver City, CA, USA) and then transferred to the micro-CT (ImTek Inc., Knoxville, TN, USA) for an 8-min scan. Parameters for CT acquisition were 70 kVp, 500 µA, and an exposure time of 480 s. A Feldkamp reconstruction algorithm was applied. PET/CT images were fused and analyzed using OsiriX Imaging software (Pixmeo, Geneva, Switzerland). Uptake in the bone marrow (femur), spleen, and thymus were normalized to muscle tissue and respectively averaged over the four mice. Imaging data are presented in Figure 8.

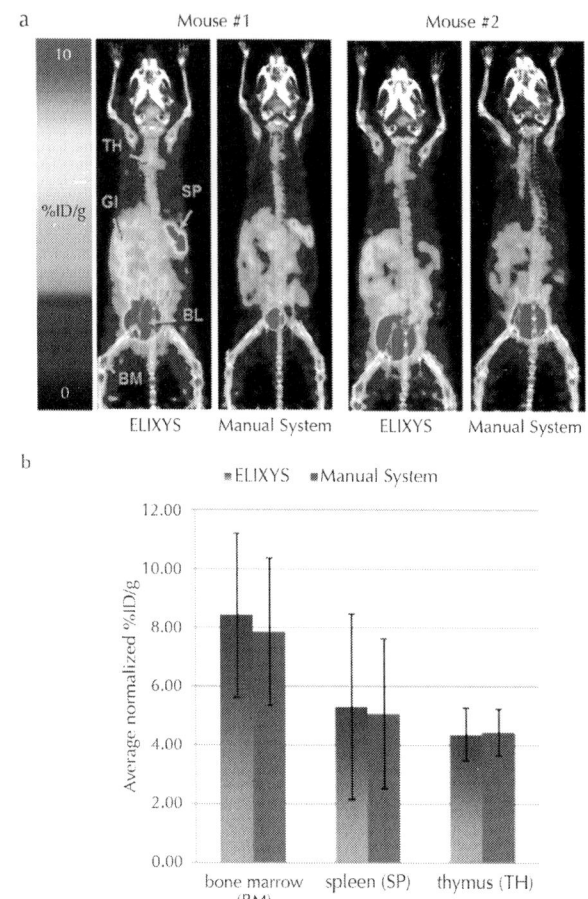

Figure 8: In vivo imaging of D-[18F]FAC. (a) PET/CT imaging of two of the four C57BL/6 mice using D-[^{18}F]FAC produced from either ELIXYS or a manually operated apparatus (TH, thymus; GI, gastrointestinal tract; SP, spleen; BL, bladder; BM, bone marrow). (b) Percent injected dose per gram (% ID/g) was normalized to muscle tissue and averaged for the four mice in each organ of interest.

All animal experiments were approved by the UCLA Animal Research Committee and were performed according to the guidelines of the Division of Laboratory Animal Medicine at UCLA.

RESULTS AND DISCUSSION

To validate the functionality of the synthesizer, the three-reactor syntheses of D-[^{18}F] FAC and L-[^{18}F] FMAU were performed. The maximum radioactivity used in the syntheses was approximately 35 GBq. Decay-corrected radiochemical yield, duration of synthesis, and specific activity are listed in Table 2. Synthesis times and yields are comparable to or better than other reports found in the literature [11, 16, 17, and 24].

Table 2: Comparison of synthesis data

Radiotracer	Reference	Radiochemical yield (%)	Duration of synthesis (min)	Specific activity (GBq/µmol)
[^{18}F]FAC	Amarasekera et al. [17]	39 ± 5 (n = 13)	~240	>37
	This work	31 ± 5 (n = 6)	165	37 to 44
[^{18}F]FMAU	Alauddin et al. [11]	20 to 30	210 to 240[a]	85.1
	Mangner et al. [24]	42.1 ± 12.1 (n = 9)	160	111[b]
	Li et al. [16]	12 ± 3 (n = 4)	150[a]	14.2 ± 1.2
	Amarasekera et al. [17]	35 ± 6 (n = 10)	~240	>37
	This work	46 ± 1 (n = 6)	170	100 to 170

Yield, synthesis duration, and specific activity of two tracers produced with the ELIXYS radiosynthesizer compared to values found in literature. [a]End point not specified, but presumably includes HPLC purification; [b]time point not specified, but presumably at time of injection.

Lazari et al.

Lazari et al. EJNMMI Research 2013 3:52, doi: 10.1186/2191-219X-3-52

The ELIXYS currently supports PET tracer production in our preclinical facility, and therefore, products were confirmed by coinjection with cold standard into the analytical HPLC, and radiochemical purity was found to be >99% for both tracers. However, numerous batches of D-[^{18}F] FAC (as well as other tracers) have been subjected to the complete set of quality assurance tests required for clinical use and have passed. These tests include residual solvent analysis via gas chromatography, Kryptofix K_{222} spot test, filter integrity test, radionuclide identity by half-life and energy, visual inspection of optical clarity, as well as pyrogenicity and sterility. We are currently updating the software for compliance with current good manufacturing practice, and the updated system will be placed in a clinical facility for production of tracers under 21 CFR 212 regulations or USP 823 guidelines.

In vivo imaging using D-[^{18}F] FAC produced on ELIXYS and that produced by the UCLA Biomedical Cyclotron Facility in a manually operated apparatus [17] showed comparable images and uptake as expected in the gastrointestinal tract (GI) and hematopoietic organs (Figure 8) [25]. Though there appears to be higher 'noise' (muscle uptake) in the images of mice injected with ELIXYS-produced D-[^{18}F] FAC, there is also higher 'signal' (organ uptake). Indeed, whole-body regions of interest excluding the tail showed that the total amount of activity was higher in the images of the mice injected with the ELIXYS-produced tracer. We suspect that there may have been variations in the injections, e.g., more tracer left in the tail in some of the images, leading to a lower amount of activity in circulation. To remove these biases, organ uptake was normalized to a region of muscle tissue that showed no significant uptake.

The disposable cassette approach has allowed for multiple tracers in addition to D-[^{18}F]FAC and L-[^{18}F]FMAU to be readily synthesized, including 2-deoxy-2-[^{18}F]fluoro-5-ethyl- -D-arabinofuranosyluracil (D-[^{18}F]FEAU) [26] as well as 2-[^{18}F]fluoro-2-deoxy-D-glucose ([^{18}F]FDG), 3-deoxy-3-[^{18}F]fluoro-L-thymidine ([^{18}F]FLT), [^{18}F]fallypride, 9-(4-[^{18}F]-fluoro-3-hydroxymethylbutyl)-guanine ([^{18}F]FHBG), and N-succinimidyl-4-[^{18}F]fluorobenzoate ([^{18}F]SFB) (results to be published separately), by simply switching cassettes and software programs. No hardware or plumbing changes were needed between productions of different tracers.

CONCLUSIONS

We have developed a new, versatile radiosynthesizer that is suitable both for reaction development and routine production of PET tracers. The ELIXYS synthesizer contains very few wetted fittings and tubing compared to other radiosynthesizers, yet through the unique use of motion to implement dynamically reconfigurable fluid paths, it is capable of diverse syntheses without hardware customization. It has been designed to be capable of diverse syntheses requiring up to three reaction vessels, high-pressure and high-temperature reactions, as well as sensitive, corrosive, and volatile reagents.

To validate this new system, the three-reactor syntheses of D-[^{18}F] FAC and L-[^{18}F] FMAU were demonstrated, and yields and synthesis times were found to be comparable to other reports. Several additional tracers of varying complexity have also been successfully synthesized on this system.

AUTHORS' CONTRIBUTIONS

ML, KMQ, SBC, GJS, MDM, and RMvD designed the synthesizer. ML, KMQ, SBC, and BM built and tested the synthesizer. HEH contributed key design elements to the synthesizer and software. ML, KMQ, SBC, MDM, and RMvD designed the software architecture and user interface. KMQ and SBC implemented and tested the software. ML, KMQ, and SBC designed and performed experiments to characterize the synthesizer. ML performed the radiosyntheses and analyses of synthesis data. JC helped perform the radiosyntheses and collect specific activity data. MEP, MDM, and RMvD conceived the project. ML and RMvD wrote the manuscript. All authors read and approved the final manuscript.

ACKNOWLEDGMENTS

This work was supported in part by the Department of Energy Office of Biological and Environmental Research (DE-SC0001249 and DE-FG02-06ER64249), the UCLA Foundation from a donation made by Ralph and Marjorie Crump for the Crump Institute for Molecular Imaging, and Sofie Biosciences, Inc. We would like to acknowledge Dan Thompson and the rest of the team at Design Catapult for their insight and creativity in tackling some of the complex engineering challenges. We greatly appreciate Dr. Saman Sadeghi and Phillip Marchis for providing D-[^{18}F] FAC from the UCLA Biomedical Cyclotron Facility; Drs. David Nathanson, Liu Wei, Caius Radu, and Johannes Czernin for generously supporting the microPET imaging; and Dr. Nagichettiar Satyamurthy for his invaluable advice.

REFERENCES

1. Phelps ME: Positron emission tomography provides molecular imaging of biological processes. PNAS 2000, 97:9226-9233.
2. Lasne M-C, Perrio C, Rouden J, Barré L, Roeda D, Dolle F, Crouzel C: Chemistry of $^+$-emitting compounds based on fluorine-18. In Edited by Krause W. Edited by Contrast Agents II. Berlin: Springer; 2002:201-258.
3. Weber WA, Figlin R: Monitoring cancer treatment with PET/CT: does it make a difference? J Nucl Med 2007, 48(Suppl 1):36S-44S.
4. Sachinidis JI, Poniger S, Tochon-Danguy HJ: Automation for optimised production of fluorine-18-labelled radiopharmaceuticals. Curr Radiopharm 2010, 3:248-253.
5. Keng PY, Esterby M, van Dam RM: Emerging technologies for decentralized production of PET tracers. In Positron Emission Tomography - Current Clinical and Research Aspects. Edited by Hsieh CH. Rijeka: InTech; 2012:153-182.

6. Coenen HH, Elsinga PH, Iwata R, Kilbourn MR, Pillai MRA, Rajan MGR, Wagner HN Jr, Zaknun JJ: Fluorine-18 radiopharmaceuticals beyond [18F]FDG for use in oncology and neurosciences. Nucl Med Biol 2010, 37:727-740.
7. Alauddin MM, Shahinian A, Gordon EM, Conti PS: Direct comparison of radiolabeled probes FMAU, FHBG, and FHPG as PET imaging agents for HSV1-tk expression in a human breast cancer model. Mol Imaging 2004, 3:76-84.
8. Chin FT, Namavari M, Levi J, Subbarayan M, Ray P, Chen X, Gambhir SS: Semiautomated radiosynthesis and biological evaluation of [18F] FEAU: a novel PET imaging agent for HSV1-tk/sr39tk reporter gene expression. Mol Imaging Biol 2007, 10:82-91.
9. Radu CG, Witte ON, Nair-Gill ED, Satyamurthy N, Shu CJ, Czernin J: Positron emission tomography probes for imaging immune activation and selected cancers. 2012. US Patent 8101740
10. Laing RE, Walter MA, Campbell DO, Herschman HR, Satyamurthy N, Phelps ME, Czernin J, Witte ON, Radu CG: Noninvasive prediction of tumor responses to gemcitabine using positron emission tomography. Proc Natl Acad Sci 2009, 106:2847-2852.
11. Alauddin MM, Conti PS, Fissekis JD: Synthesis of [18F]-labeled 2'-deoxy-2'-fluoro-5-methyl-1- -D-arabinofuranosyluracil ([18F]-FMAU). J Label Cpd Radiopharm 2002, 45:583-590.
12. Cai H, Li Z, Conti PS: The improved syntheses of 5-substituted 2'-[18F] fluoro-2'-deoxy-arabinofuranosyluracil derivatives ([18F] FAU, [18F] FEAU, [18F] FFAU, [18F] FCAU, [18F] FBAU and [18F] FIAU) using a multistep one-pot strategy. Nucl Med Biol 2011, 38:659-666.
13. Alauddin MM, Fissekis JD, Conti PS: Synthesis of [18F] ▢ labeled adenosine analogues as potential PET imaging agents. J Label Compd Radiopharm 2003, 46:805-814.
14. Anderson H, Pillarsetty N, Cantorias M, Lewis JS: Improved synthesis of 2'-deoxy-2'-[18F]-fluoro-1-[beta]-d-

arabinofuranosyl-5-iodouracil ([18F]-FIAU). Nucl Med Biol 2010, 37:439-442.

15. Paolillo V, Riese S, Gelovani JG, Alauddin MM: A fully automated synthesis of [18F]-FEAU and [18F]-FMAU using a novel dual reactor radiosynthesis module. J Label Compd Radiopharm 2009, 52:553-558.

16. Li Z, Cai H, Conti PS: Automated synthesis of 2'-deoxy-2'-[18F]fluoro-5-methyl-1- -d-arabinofuranosyluracil ([18F]-FMAU) using a one reactor radiosynthesis module. Nucl Med Biol 2011, 38:201-206.

17. Amarasekera B, Marchis PD, Bobinski KP, Radu CG, Czernin J, Barrio JR, van Dam RM:High-pressure, compact, modular radiosynthesizer for production of positron emitting biomarkers. Appl Radiat Isot 2013, 78:88-101.

18. Herman H, Flores G, Quinn K, Eddings M, Olma S, Moore MD, Ding H, Bobinski KP, Wang M, Williams D, Williams D, Shen CK, Phelps ME, van Dam RM: Plug-and-play modules for flexible radiosynthesis. Appl Radiat Isot 2013, 78:113-124.

19. Herman H, Quinn K, Esterby M, Shah GJ, Flores G, Phelps ME, Satyamurthy N, van Dam RM:Flexible radiosynthesizer capable of multi-pot high temperature and pressure reactions [abstract]. J Label Compd Radiopharm 2011, 54:s50.

20. Claggett SB, Quinn KM, Lazari M, and Moore MD, van Dam RM: Simplified programming and control of automated radiosynthesizers through unit operations. EJNMMI Research 2013, 3:53.

21. Moore MD, Quinn K, Claggett S, Lazari M, Shah GJ, Satyamurthy N, Phelps ME, van Dam RM: ARC-P HS+: a versatile radiosynthesizer for the production of PET tracers [abstract]. Philadelphia: American Association for Cancer Research; 2012. [American Association for Cancer Research Annual Meeting: 31 March–4 April 2012; Chicago]

22. Iwata R, Yamazaki S, Ido T: Automated injection of a radioactive sample for preparative HPLC with feedback control. Int J

Radiat Appl Instrum Appl Radiat Isot 1990, 41:1225-1227.
23. Berridge MS, Apana SM, Hersh JM: Teflon radiolysis as the major source of carrier in fluorine-18. J Label Compd Radiopharm 2009, 52:543-548.
24. Mangner TJ, Klecker RW, Anderson L, Shields AF: Synthesis of 2'-deoxy-2'-[18F]fluoro-[beta]-D-arabinofuranosyl nucleosides, [18F]FAU, [18F]FMAU, [18F]FBAU and [18F]FIAU, as potential PET agents for imaging cellular proliferation: synthesis of [18F]labelled FAU, FMAU, FBAU, FIAU. Nuc Med Biol 2003, 30:215-224.
25. Radu CG, Shu CJ, Nair-Gill E, Shelly SM, Barrio JR, Satyamurthy N, Phelps ME, Witte ON:Molecular imaging of lymphoid organs and immune activation using positron emission tomography with a new 18F-labeled 2'-deoxycytidine analog. Nat Med 2008, 14:783-788.
26. Shen B, Lazari M, Maraglia B, Collier L, Hammond K, Moore M, van Dam RM, Chin FT:Improved [18F]FEAU radiosynthesis on different automated radiochemistry platforms [abstract]. J Label Compd Radiopharm 2013, 56:s471.

Chapter 6

Quality of Electroless Ni-P (Nickel-Phosphorus) Coatings Applied in Oil Production Equipment with Salinity

Fernando B. Mainier, Maria P. Cindra Fonseca,
Sérgio S. M. Tavares, Juan M. Pardal

Escola de Engenharia, Universidade Federal Fluminense (UFF), Niterói, Brazil

ABSTRACT

The corrosion resistance of nickel-phosphorus (Ni-P) coatings and their mechanical properties in seawater have led investigations into the development of new technologies and the replacement of some special alloys in equipment used in oil production, such as valves, tubing, sucker rod joints, pumps, riser, manifolds and subsea Christmas trees. These studies began with Brenner and Riddel who developed, in the 1940s, formulations for Ni-P deposition on carbon steel without using an electric current. Joint deposition of nickel and phosphorus on a metallic surface (carbon steel) without applying an external current is accomplished using cathodic reduction with hydrogen (H) from a reducing agent (sodium hypophosphite) and nickel salts. To assure good performance of a Ni-P coating, the deposit quality must be inspected and evaluated during the chemical deposition process or in the end product. The recommended test parameters are: thickness, layer uniformity, hardness, adhesion, porosity, corrosion resistance and chemical composition of the nickel-phosphorus coating. The purpose of this paper was to investigate the Ni-P coating process, to evaluate the behaviour of Ni-P in a saline environment using aqueous brine (3.5% - 30% sodium chloride by mass) and to present possible defects that could compromise the coating.

INTRODUCTION

One of the traditional techniques to maintain the mechanical characteristics of a material in the manufacturing of industrial equipment, primarily using carbon steel or low alloy steels, as well as to make surfaces more resistant to abrasion and corrosion, is without doubt, by applying a specific finish, such electroless nickel-phosphorus plating (Ni-P). The Ni-P coating is deposited on carbon steel without the application of an external electrical current. This feature has therefore led to, directly or indirectly, the development of special tools and new technologies in the area of oil production in a high salinity environment associated with corrosive gases such as CO_2 and H_2S.

Historically, the process of nickel deposition was initiated in 1844 by the work of Wurtz [1,2], who discovered the reduction of Ni^{2+} ions to metallic porous nickel (Ni), and subsequently by Brenner and Riddel [3], who over one hundred years later developed formulations and practices for Ni-P deposition on carbon steel without the aid of an electric current. Studies by Duncan [4], Colaruotolo [5], Mainier et al. [6], Tallinn [7], Weil et al. [8], Mainier and Araújo [9], Delaunois et al. [10], Liu et al. [11] and Baudrand [12] have shown that the rate of growth and application of Ni-P coatings since the 1980s has increased in several industrial sectors.

The performance of this finish has led to its use in various industrial areas such as the production of pulp and paper, plastics, petrochemicals, textiles, automobiles, aeronautics, electronics and food. Also, faced with high corrosivity and the increasing challenges of obtaining petroleum under adverse conditions have driven the oil industry to seek new alternatives in the areas of materials. Ni-P coatings, due to their abrasion and corrosion resistance, have presented excellent performance, mainly in terms of valves, special tools, risers, pumps and production pipes. Figure 1 shows some of these applications in the petroleum industry.

The technical literature and patents concerning nickel deposition processes indicate a wide range of development in terms of new formulations, additives and other co-precipitations with cobalt (Co), boron (B), silicon carbide (SiC), Teflon, etc., which will provide new properties to Ni-P coatings and therefore lead to new Industrial applications.

DEPOSITION PROCESS OF NICKEL-PHOSPHORUS

The electrochemical deposition process, without the aid of an electric current, is presented in Figures 2 and 3. The process starts with the receipt of parts in quality control where they are formalized according to operational procedures that should assess the following parameters: dimensional analysis, definition and assessment of the constituent materials of the part, rules and procedures of surface preparation (chemical or mechanical cleaning for removal of oxides, oily material or grease), the type of bath to be used, temperature, speed of deposition, thickness, and the nickel/phosphorus ratio, among others.

Mechanism and Properties of Ni-P Deposition

In the conventional process of nickel electroplating, Ni^{2+} ions (present in the bath) are reduced to metallic nickel by an external electric current and deposited onto the surface of a metallic material connected to the negative pole (cathode), while the positive pole (anode) is usually made up of high purity nickel, as shown in the illustration in Figure 4. The thickness of the deposited nickel film and its properties depend on the electric current density, the voltage in volts applied, the concentration of salts, the bath temperature, pH, the nature of the base metal (cathode) and the additives used to give specific features to the nickel coating.

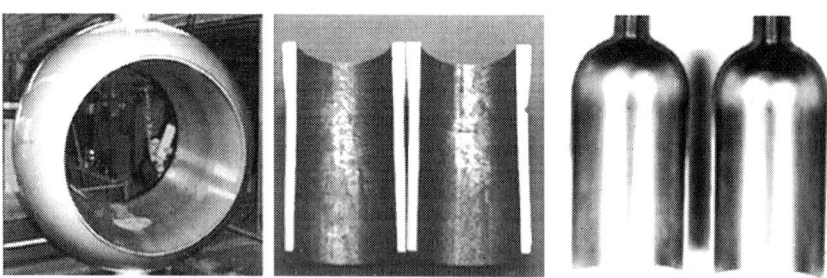

Figure 1: Ni-P coating process in pipeline valve, production pipe and the interior of a corrosive gas storage cylinder.

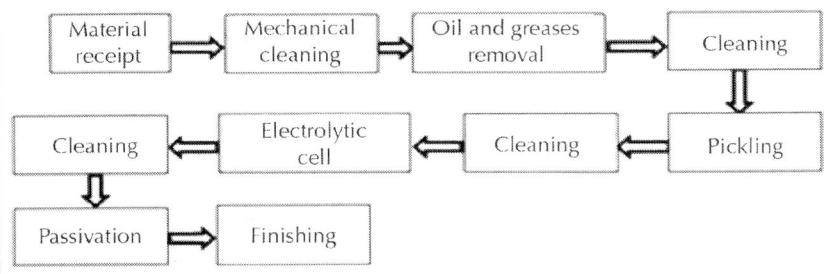

Figure 2: Deposition process.

Figure 3: Overview of the bath of nickel-phosphorus.

The first point that differentiates this process is that the deposition of autocatalytic chemical nickel plating requires no external electric current; that is, the process is self-regulated by the kinetics of the reactions involved. Continuous, uniform joint co-deposition of nickel and phosphorus is achieved by cathodic reduction with atomic hydrogen (H) produced in the bath from the hydrolysis of the reducing agent (NaH_2PO_2). The baths used in the autocatalytic process are more complex and require more control, generally, and are formulated based on nickel salts (Ni^{2+}), the reducing agent (NaH_2PO_2) and additives that control the pH, complexing and the addition of other salts to ensure the quality of the coating [13].

The required scientific and technological knowledge, as well as the reaction mechanisms to explain the deposition of Ni-P, are credited to Gould et al. [14], Duncan [4], Belinsky [15], Mallory [16] and Riedel [17], among others. It can be assumed, on the basis of these authors, that the coating deposition kinetics of Ni-P is based on the following main points and shown in Figure 5.

- Atomic hydrogen capacity formation;
- Hydrogen adsorption capacity by the metal surface;
- Reduction of Ni^{2+} ions and sodium hypophosphite;
- Co-deposition of nickel and phosphorus on the metal surface.

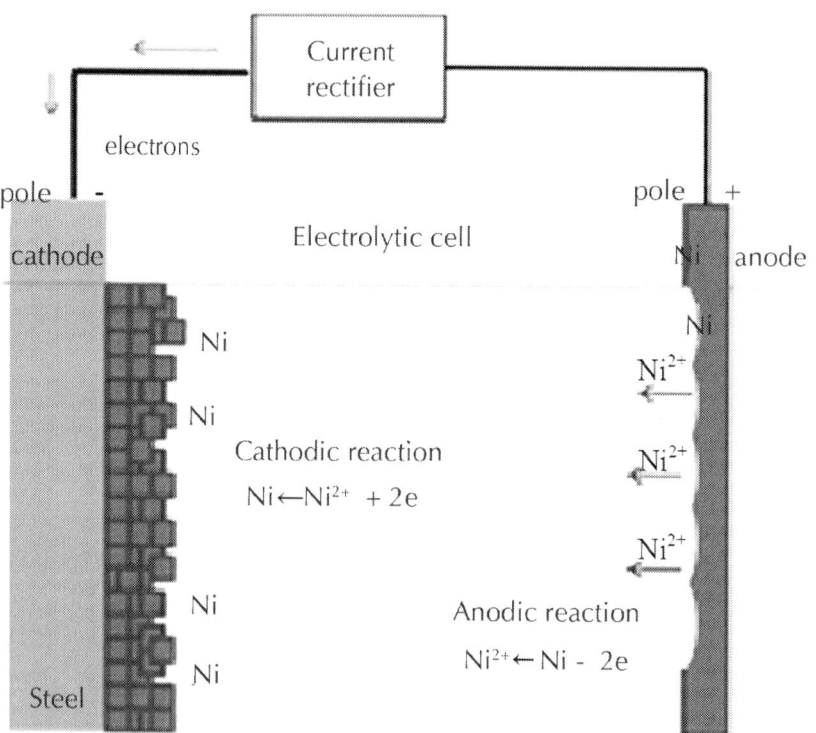

Figure 4: Illustration of electrolytic deposition of nickel on carbon steel.

Atomic hydrogen (H) formation and adsorption in the metallic surface produced by the reaction of the hypophosphite ion with water.

$H_2PO_2^- + H_2O \rightleftharpoons H^+ + H_2PO_3^{2-} + 2H$

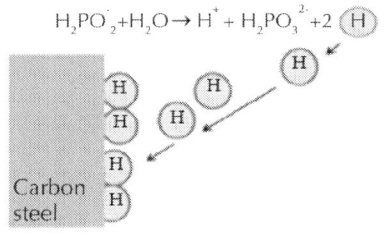

Reduction for atomic hydrogen (H) of Ni2+ ion and hypophosphite ($H_2PO_2^-$) ion, co-deposition of nickel-phosphorus (NiP) and new adsorption of atomic hydrogen on the deposit of formed NiP

$H_2PO_2^- + H \rightarrow H_2O + OH^- + P$
$Ni^{2+} + 2H \rightarrow Ni + 2H^+$

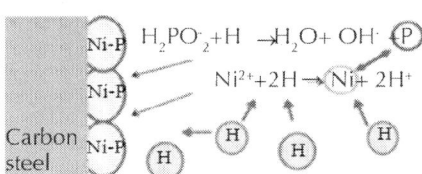

Atomic hydrogen (H) formation and adsorption in the NiP deposit produced by the reaction of the hypophosphite ion with water.

$H_2PO_2^- + H_2O \rightarrow H^+ + H_2PO_3^{2-} + 2H$

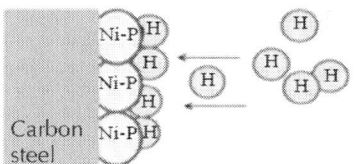

New reduction for atomic hydrogen (H) of Ni2+ ion and hypophosphite ($H_2PO_2^-$) ion, co-deposition of nickel-phosphorus (NiP) in the NiP deposit.

$H_2PO_2^- + H \rightarrow H_2O + OH^- + P$
$Ni^{2+} + 2H \rightarrow Ni + 2H^+$

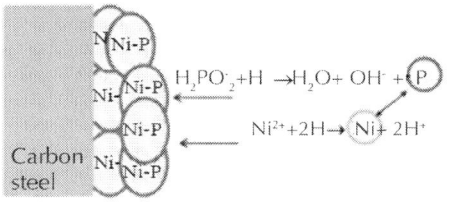

Figure 5: Mechanism of Ni-P layer formation of on carbon steel.

In the view of Etcheverry [18], the co-deposition of Ni-P has the behavior of an alloy, either crystalline or amorphous, depending on the percentage (mass %) of nickel and phosphorus. Figure 6(a), below, shows the microscopic aspects of Ni-P deposition observed by scanning electron microscopy; Figure 6(b) shows an image taken with an optical microscope.

It is important to note that parallel or unexpected reactions can occur in the bath by reducing the ability of codeposition Nl-P coating, such as:

- $H_2PO_2^- + H_2O \rightarrow H^+ + HPO_3^{2-} + H_2$—the hydrolysis of hypophosphite to form molecular hydrogen (H_2) instead of atomic hydrogen (H), which entails a decrease in the reducing power;
- $2H \rightarrow H_2$—the natural loss of atomic hydrogen (H) reduction capacity;
- $Ni^{2+} + HPO_3^{2-} \rightarrow NiHPO_3$—the possibility of nickel ion (Ni^{2+}) precipitation in the form of nickel (II) hydrogen phosphite ($NiHPO_3$), resulting in the impoverishment of the Ni^{2+} ion concentration in the bath or, if deposited in the Ni-P coating, it can make a rougher coat.

To avoid this kind of problem, it is essential that substances are present in the bath that act on complexants of Ni^{2+} ions (the preventing precipitation). For this, we used organic acid-based formulations, such as tartaric acid, malic acid, succinic acid, adipic acid, hydroxypropionic acid, etc. In addition, the concentration of hydrogen phosphite (HPO_3^{2-}) ions increases as atomic hydrogen (H) is produced. This concentration should be limited and the occurrence of hydrogen phosphite ion co-deposition along with Ni-P should be monitored as this can lead to a porous coating. Laboratory experiments have shown that the addition of activating substances such as adipic acid, succinic acid, etc. can slow the deposition speed of Ni-P and to avoid this kind of problem [17].

In this way, the deposition process is complex and, therefore, the speed of Ni-P deposition on the surface of a part, according to Riedel [17], is preferably a function of the following parameters: temperature, pH, concentration (nickel hypophosphite salts, complexants, activators, stabilizers and contaminants), surface roughness, agitation of the bath and the area/volume relationship of the electroless nickel bath.

Figure 6: (a) Scanning electron microscopy (800×); (b) Optical microscopy (200×).

According to Riedel [17], the temperature and the pH are important parameters since, as the deposition temperature and pH increase, in relation to most baths, an increased speed of deposition is favored, which can create voids or porosity in the deposited layer. The graphs presented in Figure 7 are based on baths containing 15 g/L nickel chloride and 10 g/L sodium hypophosphite to show this trend.

NI-P COATING: SPECIFICATIONS AND PROPERTIES

The following are some properties of Ni-P coatings [17,19] based on the nickel and phosphorus levels (Table 1).

For a Ni-P coating applied on carbon steel come meet the adverse conditions of the production of petroleum products, it is essential to adopt a methodology critical inspection "in situ", with due allowance for the manufacturing process of the equipment and the process of deposition based on ISO 4527 [20].

In the critical inspection of process of Ni-P deposition is important to know the various routines comprising the process itself, and must be inspected, among others, the areas of preparation (sandblasting and cleaning chemistry), the process control laboratory (qualified personnel, equipment and assessment procedures), electrochemical baths and the finishing area.

In the actual inspection of the parts should be required to evaluate the following parameters: appearance, uniformity, layer thickness, abrasion, chemical composition, adhesion, porosity and hardness. In addition, with a view to the corrosivity of petroleum, containing CO_2 and H_2S [21], corrosion tests in order to increase the security of the application and the use of Ni-P coating under adverse conditions.

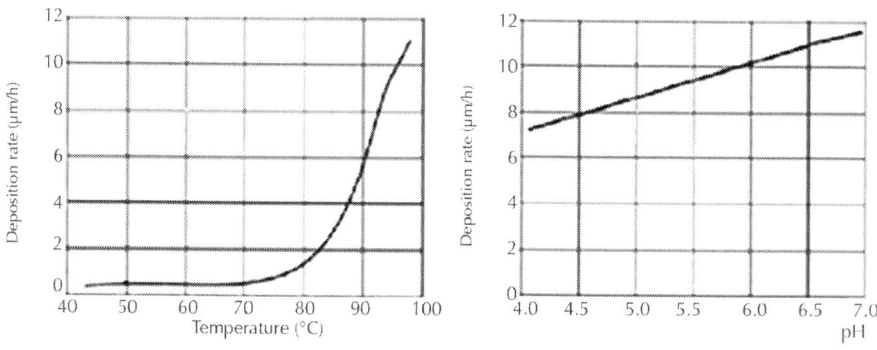

Figure 7: Speed of deposition (μm/h) of Ni-P as a function of pH and temperature.

Table 1: Properties of nickel-phosphorus coatings

Properties	Phosphorus content (%)		
	Low	Medium	High
Nickel, % (mass)	96-99	92-95	88-91
Phosphorus, % (mass)	1-4	5-8	9-12
Vickers microhardness without heat treatment, HV	650-750	500-550	450-500

Vickers microhardness with heat treatment, HV	1000-1050	900-950	850-900
Melting point, °C	1200	890	870
Density, g/cm3	8.5-8.7	8.1-8.3	7.7-7.8
Resistivity, μΩ/cm	50	70	90
Resistance to abrasion	Superior	Very good	Very good
Weldability	Good	Regular	Bad

Appearance

On the coated parts visual inspection should not show any defects such as pitting, exfoliations, bubbles, cracks, deposits or failures that may constitute an impediment on the performance of the material.

Deposit Thickness and Uniformity

To work in high aggressive environments are recommended thicknesses ranging from 75 to 125 μm. The procedure for the determination of the thickness of Ni-P deposit must be specified by the user and/or established by common agreement between the parties, and may be used microscopic methods, magnetic, ultrasonic, etc.

Optical microscopy determines the thickness and uniformity of deposition, however, this method is destructive. To solve this type of problem is allowed the use of a test coupon (testimony) representative for measurement of coating thickness. For a comparison, the micrographs of Ni-P coating applied on carbon steel show in Figure 8, the following are examples of uniformity of layer (Figure 8(a)), while in Figure 8(b) are presented defects occurring during processing of deposition.

Chemical Composition

The corrosion resistance of Ni-P coating depends on the concentration of nickel and phosphorus present in the deposited layer and increasing the phosphorus content improves the anti-corrosion protection. The ISO 4527 [20] standard presents in the Table 2 the chemical composition acceptable for Ni-P coatings. The determination of nickel and phosphorus content in the Ni-P deposit can be made using the following analytical techniques: wet chemical, atomic absorption, x-ray fluorescence, plasma, etc. For the severe conditions of industrial use the phosphorous content shall not be less than 10% by mass.

Porosity

The porosity of the electroless Ni-P coating is a most important parameter. The deposition of nickel-phosphorus should be free of porosity in order to prevent the corrosive medium contact with the base metal (carbon steel), the corrosive process begins, often through pores or failures. This problem becomes dangerous when the metal is anodic in the relation to Ni-P deposit, resulting in a galvanic cell (galvanic corrosion).

In the evaluation of Ni-P coating porosity is used the Ferroxyl Method [20,22]. This method is to place a filter paper on the piece and then applies a mixture of solution $K_3Fe(CN)_6$—potassium ferricyanide and sodium chloride on the piece during 30 seconds.

The appearance of blue points indicates the porosity and the attack on the base metal (carbon steel) as shown, then, Figure 9. The reactions: $Fe - 2e \rightarrow Fe^{2+}$ $3Fe^{2+} + 2[Fe(CN)_6]^{3-} \rightarrow Fe_3[Fe(CN)_6]_2$ (blue points).

correspond, to attack the carbon steel and the blue color development on the filter paper placed on the piece or test coupon.

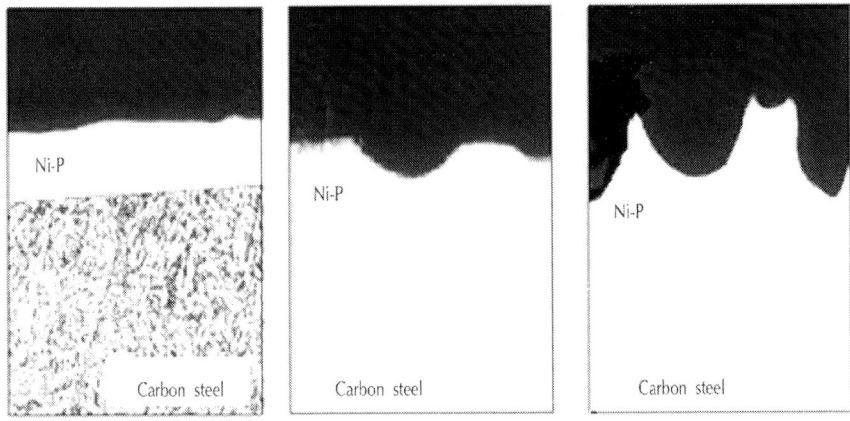

Figure 8: Optic micrographs of Ni-P coating applied on carbon steel: (a) Uniform; (b) Defective layers.

Table 2: Chemical composition of Ni-P deposit [20]

Elements	Chemical composition (% mass)		
	Minimum	Maximum	Typical
Nickel	85	98	88-95
Phosphorus	2	15	5-12
Other (Al, As, B, Bi, C, Cd, Co, Cr, Cu, Fe, Mn, Nb, Pb, S, Sb, Se ,Si, Sn, V, Zn)	0	2.0	0.05

Microhardness

The hardness of Ni-P deposit indicates whether part or not specific heat treatment suffered and usually ranges from 500 to 580 HV (Vickers microhardness). After heat treatment depending on the exposure time and temperature the hardness can vary from 600 to 1100 HV. The heat treatment applied to Ni-P coatings shall comply with the directions of the rules in order to minimize the occurrence of cracks or fissures [20,23]. Agarwala & Agarwala [24] show that the types of baths determine the relationships of the nickel and phosphorus levels in the deposit and the increase of the levels of phosphorus reduces the hardness with and without heat treatment as shown in the Table 3.

Adhesion

The adhesion of electroless Ni-P to non-conductors is dependent on mechanical keying with associated Van des Waals force. Ni P coatings have good adherence to carbon steel due to the fact that the cohesion of the forces on the metal-base film are often in superior of 140 MPa (20,000 psi). The evaluation of this adhesion is essential and often so that the tests are acceptable, it is indispensable to carry out various tests. These tests are usually comparative, qualitative and quantitative tests are based on an aluminum disk collage on the metallic surface and applying a continuous tension force in aluminum disk until the breakup. However, its implementation depends on the geometry of the piece and its use. Adhesion tests are based on standards: ASTM B 733 [25] and ISO 4527 [20].

Abrasion

Resistance to abrasion is directly related to the phosphorous content, heat treatment and adherence to the surface of the base metal. In general, the increase of the levels of phosphorus and the increase of hardness provided by heat treatment increases the abrasion resistance. Abrasion tests are specified and depending on the specific use of electroless Ni-P coating as recommends ISO 4527 [20].

Corrosion

In evaluating the performance of Ni-P coating in different aggressive media are used in laboratory testing and field. In the case of production of oil this Ni-P coating has proven quite attractive, with a view to its good resistance to many corrosive media such as corrosive gases (CO_2, H_2S) and high water salinity, contaminants commonly found in petroleum. However, it is essential to establish two factors are important in corrosion resistance: the phosphorous content which should be more than 10 % (mass) and the thickness shall not be less than 75 mm.

Electroless Ni-P coating are not recommended for some corrosive media containing chloride based compounds and ferric sulfate, nitrates, nitrites and ammonium compounds. Laboratory testing to evaluate the performance of electroless Ni-P coating in petroleum with high aggressiveness can be static or dynamic.

The high aggressiveness can be represented by mixes with salt water and pressurized using mixtures of CO_2, H_2S. The corrosion test can be carried out in a pressurized cell to 70 kgf/cm². The temperature can range from 25°C to 80°C and exposure time of 240 hours continuous hours.

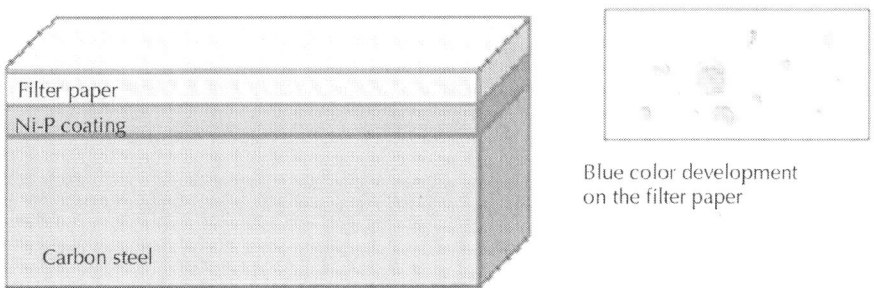

Blue color development on the filter paper

Figure 9: Ferroxyl test applied to determine the porosity of Ni-P deposit on carbon steel.

Table 3: Deposit hardness of Ni-P-carbon steel [24]

Phosphorus, % (mass)	Vickers microhardness (HV)	Vickers microhardness with heat treatment, HV
2-3	700	1000
6-9	550	920
10-12	510	880

CORROSION TESTING OF ELECTROLESS NI-P COATING

The carbon steel coupons coated Ni-P, heavy, previously were placed in pressurized containers (70 kgf/cm^2) with 300 mL containing sodium chloride solutions. Sodium chloride solutions used in the test were, respectively, of 3.5%, 10%, 20% and 30% by weight. The tests were run temperature of 25°C and for 250 hours of total immersion. The properties of coupons with Ni-P coating are presented below in Table 4.

After the end of the coupons were washed with running water, alcohol and dried with hot air and some coupons were cut to evaluate by optical metallography the corrosive effects of the salt solution.

The tests performed with Ni-P coated coupons did not show noticeable losses in mass and the analysis by optical microscopy did not find attacks on coupons, i.e. it can be considered that the corrosion rates are void.

CONCERNS OR PRECAUTIONS REGARDING NICKEL-PHOSPHORUS COATING APPLIED ON CARBON STEEL AND USED IN EXTREME CONDITIONS

Generating ideas of the Principles of Precaution, probably born in the 1970 with the fledgling company's concerns with the ethics, with the environment, with the deterioration of water resources, with the risk of contamination, with the uncertainties of processed foods and the insecurities of the technological applications in various industrial segments.

The term precaution, in the vision of sustainability and of triad industrial safety-environment-workers' health and their descendants, should mean taking measures to protect human health and the environment from possible damage that may happen. This means that the processes must be safe to avoid or minimize possible disasters, however, in the international area; there are different views that caution has different interpretations, especially when the more industrialized countries impose products and/or obsolete technologies to less developed countries.

Table 4: Properties of Ni-P coated coupons

Properties	Evaluation
Phosphorus, % (mass)	10.5
Thickness (μm)	75
Porosity	Free
Apparent defects	Free
Adhesion	Very good
Vickers microhardness (HV)	600

In the view of Mainier [26] systems productive, knowledgeable of the risks of industrial manufacturing processes and seeming not to care about the present and not the future, continue to exert strong pressure on the environment, imposing or masking obsolete technologies that include waste, packaging, recycling and toxic waste, themes that often get confused or are linked.

On the interests and economic philosophies and large industrial manufacturing complexes and industrialized countries become fellow-agents of a policy of mutual interest and, in many situations, against the interests of the man himself. In this optic, spills, leaks and contamination with large environmental impact have taken place.

Under the technical point of view can be worrying speed of deposition of Ni-P coating for high thickness in relation to the final cost of the coated part. For example, hypothetically, assuming a thickness 75 μm and with reference to three electrochemical baths, respectively, with 3 µm/h, 10 µm/h and 25 µm/h minimum operating time for these operations would be, respectively, 25 h, 7.5 h and 3 h.

In a simplistic evaluation the first case will probably have a longer time, forcing the continuous control and systematic and will have higher quality of landfill although the cost is higher when compared to other procedures. At a high speed the possibility of porosity and faults can occur with greater intensity when compared with low layer deposition speed of Ni-P.

Another point that also needs to be evaluated is the life cycle of the piece and its implications in the project as a whole when compared to application of a coating with other leagues more massive and noble and give greater security to the investment.

CONCLUSIONS

On the basis of the facts stated it can be concluded that:
- To ensure the good performance of electroless Ni-P coating is necessary to know the process of electrochemical deposition without external current and secure inspection philosophies during processing and in the final product.
- The quality of Ni-P coating must be supported for constant qualification standards and national and international procedure.
- It is essential to analyze the speed of deposition of coatings on the basis of cost and the possibility of defects to be used in extreme conditions.
- Laboratory testing showed excellent performance for Ni-P coatings with 75 µm in salt solutions of sodium chloride 3.5% to 30% by mass.

REFERENCES

1. A. C. R. Wurtz, "On Copper Hydride," Hebdomadaires des Séances de l'Académie des Sciences, Vol. 18, 1844, pp. 702-704.
2. A. C. R. Wurtz, "On Copper Hydride," Hebdomadaires des Séances de l'Académie des Sciences, Vol. 21, 1845, p. 149.
3. A. Brenner and G. E. Riddel, "Nickel Plating on Steel by Chemical Reduction," Journal of Research of the National Bureau of Standards, Vol. 37, No. 1, 1946, pp. 31- 34. http://dx.doi.org/10.6028/jres.037.019
4. R. N. Duncan, "Performance of Electroless Nickel Coated Steel in Oil Field Environments," Material Performance, Vol. 21, 1983, pp. 28-34.
5. J. F. Colaruoto, B. V. Tilak and R. S. Jasinki, "Corrosion Charactheristcs of Electroless Nickel Coating of Oil Field Environments," Proceedings of Electroless Nickel Conference IV, Chicago, 22-24 April 1985.
6. F. B. Mainier, I. M. R. A. Brüning and E. F. Pamplona, "Desenvolvimento de Recipientes para Acondicionamento de Gás Natural Contendo Gases Corrosivos," In: VI Encontro Brasileiro de Tratamento de Superfícies (EBRAT-1989), ABTS, São Paulo, 1987, pp. 66-81.
7. V. T. Talinn, "In the World of Electroless Nickel," Finishing, Vol. 12, 1988, p. 26.
8. R. Weil, J. H. Lee and K. Parker, "Comparison of Some Mechanical and Corrosion Properties of Electroless and Electroplated Nickel Phosphorus Alloys," Plating and Surface Finishing, Vol. 76, 1989, pp. 62-66.
9. F. B. Mainier and M. M. Araújo, "On the Effect of the Electroless Nickel-Phosphorus (Ni-P) Coating Defects on the Performance of This Type of Coating in Oilfield Environments," SPE Advanced Technology Series, Vol. 2, No. 1, 1994, pp. 63-67.

10. F. Delaunois, J. P. Petitjean, P. Lienard and M. JacobDuliere, "Autocatalytic Electroless Nickel-Boron Plating on Light Alloys," Surface and Coatings Technology, Vol. 124, No. 2-3, 2000, pp. 201-209.
11. X. Liu, J.-Q. Gao and W.-B. Hu, "Application of Electroless Ni-P Alloys in Electronic Industry," Plating & Finishing, Vol. 28, No. 1, 2006, pp. 30-34.
12. D. Baudrand, "Adhesion of Electroless Nickel Deposits to Aluminum Alloys—We Now Have a Better Understanding of the Factors Influencing Adhesion," Products Finishing, Vol. 63, No. 10, 2009, pp. 80-87.
13. ASTM B-656, "Auto Catalytic (Electroless) Nickel-Phosphorus Deposition on Metals for Engineering Use," American Society for Testing and Materials, West Conshohocken, 1992.
14. A. Gould, P. J. Boden and S. J. Harris, "Phosphorus Distribution in Electroless Nickel Deposits," Surface Technology, Vol. 12, No. 1, 1981, pp. 93-102.http://dx.doi.org/10.1016/0376-4583(81)90140-0
15. J. Bielinski, "The Role of Buffers and Complex Formers in Electroless Nickel Plating," Oberflache Surface, Vol. 25, No. 12, 1984, pp. 423-429.
16. G. O. Mallory and J. B. Hadju, "Electroless Plating: Fundamentals & Applications," Cambridge University Press, Cambridge, 1990.
17. W. Riedel, "Electroless Nickel Plating," Redwood Press Limited, Liverpool, 1991.
18. B. Etcheverry, "Adhérence, Mécanique et Tribologie des Revêtements Composites NiP—Talc Multifonctionnels à Empreinte Écologique Réduite," Institut National Polytechnique de Toulouse, Toulouse, 2006.
19. R. P. Tracy and G. J. Shawham, "Pratical Guide to Using Ni-P Electroless Nickel Coatings," Materials Performance, Vol. 29, No. 7, 1990, pp. 65-70.

20. ISO-4527, "Autocatalytic Nickel-Phosphorus Coatings, Specification and Test Method," International Organization for Standardization, Geneva, 1987.
21. F. B. Mainier, G. C. Sandres and R. J. Mainier, "Integrated Management System for In-House Control of Accidental Hydrogen Sulfide Leaks in Oil Refineries," International Journal of Science and Advanced Technology, Vol. 2, No. 9, 2012, pp. 76-84.
22. ASTM B-656, "Auto Catalytic (Electroless) Nickel-Phosphorus Deposition on Metals for Engineering Use", American Society for Testing and Materials, West Conshohocken, 1992.
23. ASTM B-578, "Measurements of Micro Hardness of Electroplated Coating," American Society for Testing and Materials, West Conshohocken, 1992.
24. R. C. Agarwala and V. Agarwala, "Electroless Alloy/ Composite Coatings: A Review," Sadhana, Vol. 28, No. 3-4, 2003, pp. 475-493.
25. ASTM B-733, "Standards Specifications for Autocatalytic Nickel-Phosphorus Coatings on Metals," American Society for Testing and Materials, West Conshohocken, 1992.
26. F. B. Mainier, "Uma Visão Crítica das Rotas Industriais de Fabricação de Produtos Químicos Utilizados nos Tratamentos de Água," Congresso de Equipamento e Automação da Indústria Química, Associação Brasileira da Indústria Química (ABQUIM), São Paulo, 1999.

Chapter 7

Influence of the Chemical Composition of Completion Fluids on the Propagation of Electromagnetic Waves within Oil Wells

Alexandre Ashade Lassance Cunha, Marco Aurélio Pacheco, and José Ricardo Bergmann

Departamento de Engenharia Elétrica, Pontifícia Universidade Católica do Rio de Janeiro, Rio de Janeiro, Brazil

ABSTRACT

The propagation of electromagnetic waves in the annular region of oil wells was studied. The present study aims to analyse the propagation attenuation along the well, as well as the input impedance determined by a source placed near the wellhead. A coaxial waveguide model was adopted with heterogeneous dielectrics and losses. First, a wave equation solution for the waveguide is presented, assuming a homogeneous medium with losses, by solving the equation in cylindrical coordinates using the vector potential technique. An uncertainty analysis model is then developed to model the heterogeneous characteristics of the medium. Monte Carlo simulations were performed with the created model using data gathered from the literature. The results of the simulations indicate that propagation in the transverse electromagnetic mode has the smallest attenuation and that for depths of up to 4000 m, there is an attenuation of less than 52 dB. Furthermore, the input impedance ranges from 10 Ω to 10 kΩ because of the uncertainties involved in the problem in question.

INTRODUCTION

Oil wells today are extremely complex and demand very expensive maintenance [1]. Modern drilling techniques can reach a few kilometres in depth, and the costs of a floating platform at open sea can cost billions of Brazilian Reais. Thus, it is critical to determine the internal conditions of the well, and indicators such as temperature, pressure and salinity must be constantly monitored [2].

Reliable telemetry by cables is very difficult to obtain due to the extreme conditions of the internal environment of an oil extraction well. As an example, there is continuous abrasion by sand and dirt that are carried by fluid flow. For this reason, the cables are periodically damaged and require replacements. Such replacements hinder the oil extraction process and, moreover, increase the costs due to the large amount of cables needed over the life of a single well.

The most obvious alternative to cabling is the adoption of wireless telemetry. However, the amount of electricity required to power wireless communication is not practical: power cables are needed because the use of batteries would lead to frequent maintenance, as they need to be recharged [1]. It is obvious, therefore, that the objective (and greater challenge) is to build a system without wires or batteries; that is, the downhole sensors must be powered and communicate with the base without the use of cables.

Several approaches are possible for wireless telemetry. One possibility would be the use of signal transmission techniques using a magnetic field [3]. In this method, a coil that is capable of inducing alternating current through the production pipe is used. This coil has sufficient intensity to transmit both power and the signal itself between the sensor and the base [4]. Through a second coil positioned at the sensor location, it is possible to recover the signal and power needed to feed and produce bilateral communication. However, the technique has a weakness: the production tube is not a transformer core, i.e., it is not designed to minimise the magnetic flux losses. Its hollow characteristic confers much loss by parasitic Foucault currents concomitantly with hysteresis losses, the cause of which is due to the inadequacy of the production tube material for magnetic purposes [5]. Therefore, for long distances, on the order of 3000 m or more, the method is unfeasible.

Another possible approach would be to analyse the oil well from the perspective of a coaxial propagation structure, which is formed by a conductor tube with a perfect centre (production tube) and a cylindrical shell that is coaxial with the tube and is also a perfect conductor. This approach makes it possible to conduct the analysis as it would be performed on a coaxial cable, and therefore, propagation occurs in the transverse electromagnetic mode [6]. This approach, however, has been studied without considering fluids of significant conductivity or possible uncertainties caused by temperature and electrical parameter variations of the fluid that fills the well.

The modelling difficulty concerns the fluid that fills the well. This fluid is heterogeneous because it exhibits significant variation in electric permittivity along the well depth. Furthermore, the actual fluid concentration varies from well to well, and at times within the well itself, which is another source of heterogeneity. In this context, the present article proposes to study the electromagnetic propagation inside the annular area of oil wells. The goal is to develop a model that permits an approximation of the behaviour of the TEM propagation mode in the coaxial structure with losses and the previously cited heterogeneous characteristics. The first step is the deduction of the electric and magnetic field equations within the well, assuming a typical well structure. Thereafter, statistical models of the parameters that compose the medium are developed, considering their temperature variations and propagation frequency. Finally, the study uses Monte Carlo analysis to illustrate how the propagation attenuation behaves and to analyse the input impedance of the coaxial waveguide as a function of position.

MODELLING

Well Definition

The well will be modelled according to Figure 1. This figure depicts a coaxial guide bounded by perfect conductor metals with a homogeneous dielectric and losses. The upper extremity is completely closed by a perfect conductor metal to model the valve and metal duct assembly present in a wellhead. The bottom of the well is composed of concrete; however, this fact is neglected, and the well is assumed to behave as a semi-infinite waveguide.

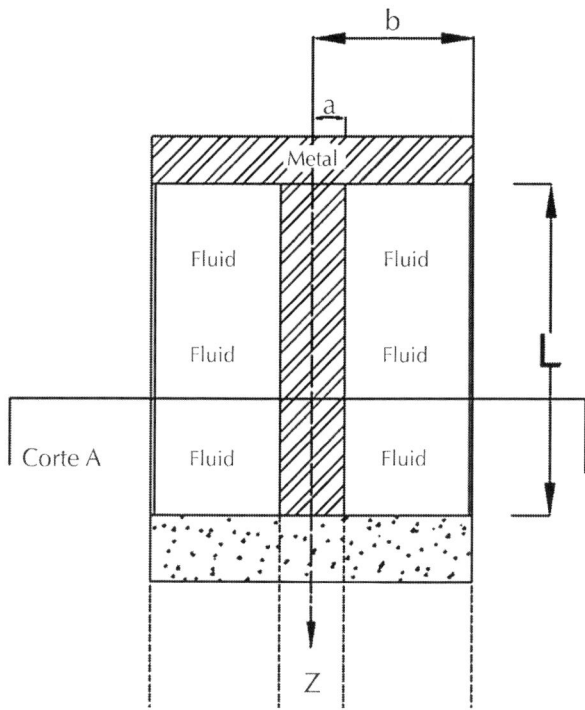

Figure 1: Adopted well structure. A semi-infinite well is assumed and is formed by two concentric cylinders and a metal block covering the top of the waveguide.

Deduction of the Propagation Equations

Modelling is performed using Maxwell's equations and the concept of vector potentials. The Maxwell's equations used are Equations (1) and (2):

$$\nabla \times E = -M_i - j\omega B \tag{1}$$

$$\nabla \times H = J_i + \sigma_e E + j\omega\varepsilon' E \tag{2}$$

The real number constant ε' is the real part of the complex electrical permittivity of the medium. Furthermore, the complex constant $_e$ is called effective conductivity and it represents the linear relation of conduction current and electric field on the medium.

Consider a medium with no source of magnetic charge $(\nabla \cdot B = 0)$. Because $\nabla \cdot \nabla \times = 0$, the curl of A can be defined as

$$\nabla \times A \equiv B = \mu H \qquad (3)$$

Substituting 3 in 1 and considering $M_i = 0$ (Induced magnetic flux equal to zero), we obtain:

$$\nabla \times E = -\nabla \times j\omega A \qquad (4)$$

Or:

$$\nabla \times (E + j\omega A) = 0 \qquad (5)$$

Since $\nabla \times \nabla = 0$, an electric scalar potential g_e can be defined such that:

$$E + j\omega A \equiv -\nabla g_e \qquad (6)$$

Applying the identity $\nabla \times \nabla \times = \nabla(\nabla \cdot) - \nabla^2$ in Equation (3) and using Equation (2), we obtain:

$$-\nabla \nabla \cdot A + \nabla^2 A$$
$$= -\mu J_i + \mu(\sigma_e + j\omega \varepsilon')(\nabla g_e + j\omega A)$$
$$\sigma_e \equiv \omega \varepsilon'' + \sigma_s \qquad (7)$$

In Equation (7), the parameter σ_e is the effective conductivity of the medium, while σ_s represents the static conductivity, that is, the conductivity when frequency is zero. At this moment, we are in a position to define $\nabla \cdot A$. To simplify Equation (7), we use the gauge:

$$\nabla \cdot A \equiv -\frac{\gamma^2}{j\omega} g_e \qquad (8)$$

Where $\gamma_1^2 \equiv j\omega\mu(\sigma_e + j\omega\varepsilon')$ is the complex constant of propagation of the medium? This step simplifies the equation for the electrical potential, resulting in:

$$\nabla^2 A = -\mu J_i + \gamma^2 A \qquad (9)$$

For homogeneous media, Equation (9), when solved, determines the electric vector potential in the medium, which can be used to obtain the equation for the electric field as a function of the electric vector potential:

$$E = \frac{j\omega}{\gamma^2} \nabla \nabla \cdot A - j\omega A \qquad (10)$$

We seek the solution in transverse electromagnetic mode (TEM), which is generated using $A = \hat{a}_z A_z$ with the restriction $E_z = 0$, which leads to the following fields in cylindrical coordinates:

$$E_\rho = \frac{j\omega}{\gamma_1} \frac{A}{\rho} \sinh(\gamma_1 z) \qquad (11)$$

$$H_\phi = -\frac{1}{\mu} \frac{A}{\rho} \cosh(\gamma_1 z) \qquad (12)$$

$$\gamma_1^2 \equiv j\omega\mu(\sigma_e + j\omega\varepsilon') \qquad (13)$$

Note that because $E_\Phi = E_z = 0$, the boundary conditions that require a null tangential component in the metal extremities are already met. As the well has a metal structure at one of the extremities, one must also ensure that $E_\rho(z = 0) = 0$. Note, however, that this condition is also already guaranteed.

The above solution represents a propagation model in TEM mode in a coaxial medium with losses inherent to the medium inside. However, it is important to remember that the solution was obtained by assuming a homogeneous propagation medium.

Statistical Modelling of the Medium Constituent Parameters

The propagation medium in the well consists of an oilbased dielectric fluid composed of water, oil and salts (typically $CaCl_2$) [7]. This fluid is the centre of all electromagnetic propagation, and therefore, its electrical characteristics must be examined in detail. A study over the range of 1 MHz to 100 MHz was previously conducted [7], revealing significantly variable behaviour based on the chemical composition of the fluid.

To model the variation of conductivity with frequency, the effective conductivity concept is applied [8] using the following formula:

$$\sigma_e \equiv \sigma_s + \omega \varepsilon'' \qquad (14)$$

From the experimental curves obtained in [8], the parameters σ_s and ε'' can be calculated using a least squares method.

An analogous model can be created to model the variation in frequency of the real relative permittivity with the frequency. With

$$\varepsilon' \equiv \varepsilon_s + K\omega \qquad (15)$$

and using the experimental curves in [8], it is possible to estimate the parameters and k using a least squares approach.

In addition to the variation in frequency, variation in the medium ε_s constituent parameter can also be observed in temperature. Such variation cannot be neglected because inside a 5000 m deep well, it is impossible to ensure that the temperature is uniform along its entire length. This fact, therefore, characterises non-homogeneity along the z direction.

To circumvent the situation, a model that assumes an average temperature in the medium is adopted. This temperature, in turn, is considered constant throughout the well, which implies a homogeneous medium. Thus, the influence of the variation of this average temperature on signal propagation in the well can be analysed, assuming a valid range of average temperatures.

Mathematically, a coefficient of correction in temperature is defined as the ratio between the temperature value in question and the value of the reference temperature, here defined as 25°C. Thus,

$$\theta \equiv \frac{\sigma_e(\theta)}{\sigma_e(25°C)} \quad (16)$$

$$\theta \approx a_1\theta^2 + a_2\theta + a_3 \quad (17)$$

$$\sigma_e = \sigma_e(25°C) \times \theta \quad (18)$$

The quadratic form was selected to present the best interpolating results using the experimental data from [8].

Similarly, there is increasing variation in the relative permittivity according to the temperature, as demonstrated by [7]. Again, it is used a quadratic model similar to the effective conductivity variation in temperature:

$$\theta_\varepsilon \equiv \frac{\varepsilon'(\theta)}{\varepsilon'(25°C)} \quad (19)$$

$$\theta_\varepsilon \approx a_1\theta^2 + a_2\theta + a_3 \quad (20)$$

$$\varepsilon' = \varepsilon'(25°C) \times \theta_\varepsilon \quad (21)$$

EXPERIMENTS AND RESULTS

In all the experiments, a well with a length of 5000 m, an inner radius of 0.05 m and an outer radius of 0.1 m is assumed. All the Monte Carlo analyses were conducted with at least 1 million samples. The objective of the experiments was to obtain graphs and numerical values for the input impedance of the "oil well" waveguide and to analyse the propagation loss in the medium for 5000 m of depth.

All analyses were conducted over the range of 1 MHz to 100 MHz. For the other free parameters, the modelling used random variables whose distribution was selected to reflect their most common values. Table 1 summarises the values selected for each free parameter of the previously described model and their respective distributions. For simplicity, when the random variable has simple and obvious domain restrictions, a uniform distribution was selected; otherwise, a normal distribution was used. Furthermore, statistical independence was assumed between the variables.

First, an experiment was conducted to calculate the attenuation in the well. The attenuation is directly dependent on the real part of the constant of propagation, whose square is defined as

$$\gamma_1^2 \equiv j\omega\mu(\sigma_e + j\omega\varepsilon').$$

The models proposed in Section 2.3 were used for the conductivity and electric permittivity, and the relative magnetic permeability was assumed to be equal to 1.

The Figure 2 presents a boxplot of the constant of attenuation for the various frequencies between 1 and 100 MHz. Values that appear in the graph as outliers are from highly unlikely combinations of the random variables of the problem and are most likely not physically feasible and, therefore, must be disregarded. The height of the boxes represents the range of values most likely to be observed, and its average point is a good approximation of the average. The boxplot of Figure 3demonstrates that the attenuation constant increases with an increase in frequency, as expected. Furthermore, note that the deviation of the coefficient increases with the elevation of the propagation frequency, which makes the system design even more difficult. Thus, it is clear that lower frequentcies are desirable from the point of view of signal attenuation.

Table 1: Random variables used for modelling uncertainties. For simplicity, when the random variable has simple and obvious domain restrictions, a uniform distribution was selected; otherwise, a normal distribution was used

Variable	Average/a	Deviation/b	Distribution
σ_s (S/m)	0	4.0×10^{-5}	Uniform
ε^s (F/m)	3	12	Uniform

ε″ (F/m)	1.0 x 10⁻¹³	3.0 x 10⁻¹²	Uniform
k (F/Hz·m)	2.5 x 10⁻⁹	5.0 x 10⁻¹⁰	Normal
θmed (°C)	40	5	Normal
a¹ (permissivity)	9.0 x 10⁻⁴	2.0 x 10⁻⁴	Normal
a² (permissivity)	-3.0 x 10⁻¹²	6.0 x 10⁻³	Normal
a³ (permissivity)	1.2	0.1	Normal
a¹ (permissivity)	1.0 x 10⁻⁵	2.0 x 10⁻⁶	Normal
a² (permissivity)	0.0	5.0 x 10⁻⁴	Normal
a³ (permissivity)	1.0	1.0 x 10⁻²	Normal

Using the TEM mode equations presented in Equation (2), it is clear that the wave propagating in the direction of the bottom of the well has power attenuation given by $A = \exp(\alpha^2 L^2)$. Using the statistical analysis of the coefficient of attenuation, Table 2 was generated.

The values from the table represent the upper limit of attenuation for 95% of the cases. Thus, for example, for a depth of 1000 m, P (A ≤ 13 dB) = 0.95. The results presented in Table 2 do not agree with the propagation for L = 5000 m of depth because of the high-energy attenuation (130.3 dB). However, it is essential to note that the table represents an upper limit of attenuation, taking 95% of the possible combinations of propagation medium and an average temperature. Moreover, as the table demonstrates, propagations up to 2000 m depth are highly acceptable, as an attenuation of up to 52.1 dB is observed in current communication technology. Even at greater depths, the possibility of communication cannot be excluded.

The second experiment was performed to analyse the input impedance of the coaxial waveguide as a function of the excitation source position. The experiment was conducted at 1 MHz.

Setting the impedance at any point as the ratio E_ρ / H_ϕ, we obtain

$$Z = Z_0 \tanh(\gamma_1 z) \tag{22}$$

$$-L < z \leq 0 \tag{23}$$

$$Z_0 = -\mu \frac{j\omega}{\gamma_1} \tag{24}$$

The position selected for analysis was at 1/4 wavelength from the wellhead, that is, $Z = \lambda/4$. However, the wavelength itself is a random variable because $\lambda = 2*\pi/\beta$ and $\beta = \text{Im}\{\gamma\}$ is a random variable. Monte Carlo analysis revealed that λ varies between 90 m and 170 m for 90% of the cases, leading to the selection of position $Z = ((90+170)*1/2)/4 = 32.5\text{m}$.

Table 2: Estimated attenuation of electromagnectic waves propagating in the annular region of the well

L (m)	Electric field attenuation (dB)	Power attenuation (dB)
1000	13.0	26.1
2000	26.1	52.1
3000	39.1	78.2
4000	52.1	104.2
5000	65.1	130.3
6000	78.2	156.3
7000	91.2	182.4

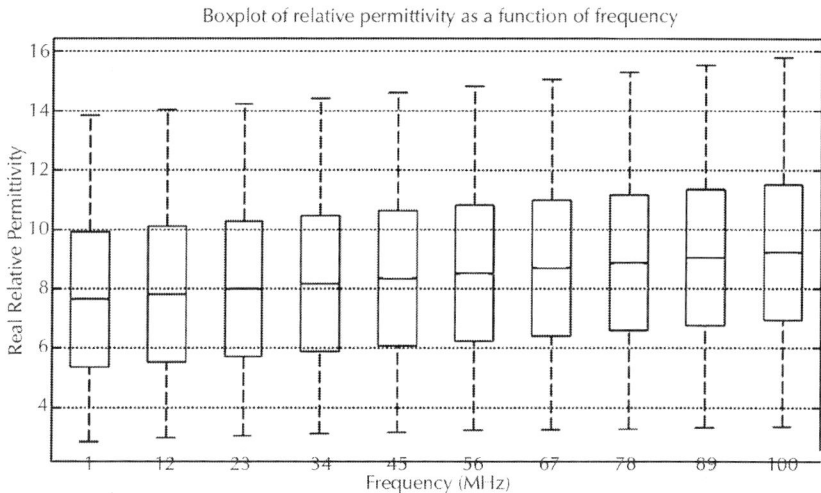

Figure 2: Boxplot of relative permittivity as a function of frequency. The horizontal axis represent the frequency of propagation in MHz, while the vertical axis represent the actual value of the electrical permittivity.

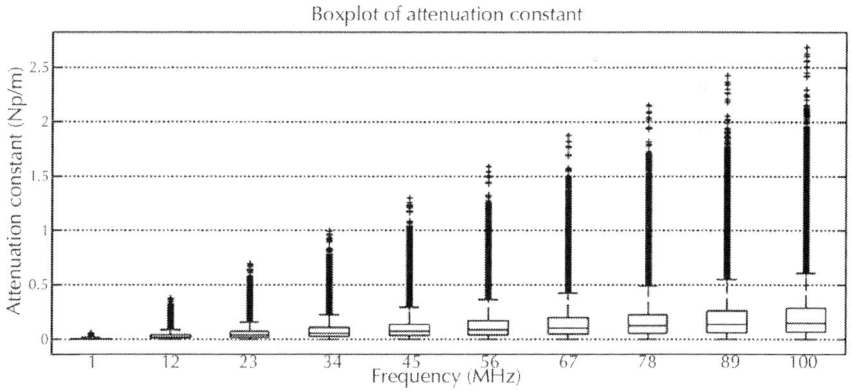

Figure 3: Boxplot of attenuation constant as a function of frequency.

Using Monte Carlo analysis again, it can be observed in Figure 4 that the input resistance (real part of the impedance) of the system exhibits great variation, ranging from 10 Ω to 1.0 kΩ for 90% of the cases. This variation is due to the variation of the relative permittivity in the medium. Therefore, it is necessary to design a generator circuit that provides a good match for a wide range of input impedances, i.e., the generator/receiver circuit must lose as little power as possible by reflection.

CONCLUSIONS

The present study focused on analysis of the electromagnetic propagation inside the annular area of oil wells, assuming the interior medium is composed of dielectrics with significant conductivity. The well behaviour was evaluated with respect to the input impedance and propagation attenuation.

To quantify the well behaviour with respect to electromagnetic propagation, a coaxial waveguide model was developed, modelling the wellhead as a metal cap that completely closes one extremity of the waveguide. First, we solved the wave equation for a homogeneous cylindrical coaxial waveguide, resulting in an analytical model. Then, an uncertainty model was adopted to approximate the heterogeneous and imprecise characteristics of the fluid that functions as a dielectric inside the well.

Using this model, two experiments were developed, both by simulation. The first aimed at analysing the input impedance observed by a source positioned at 1/4 of the wavelength from the wellhead, while the other aimed at analysing the attenuation as a function of length or well depth.

Due to the uncertainties present, the input impedance was observed to vary from 10 Ω to 1 kΩ for the frequency of 1 MHz. The conclusion was obtained by Monte Carlo simulation and was applied to the expression derived for the waveguide input impedance.

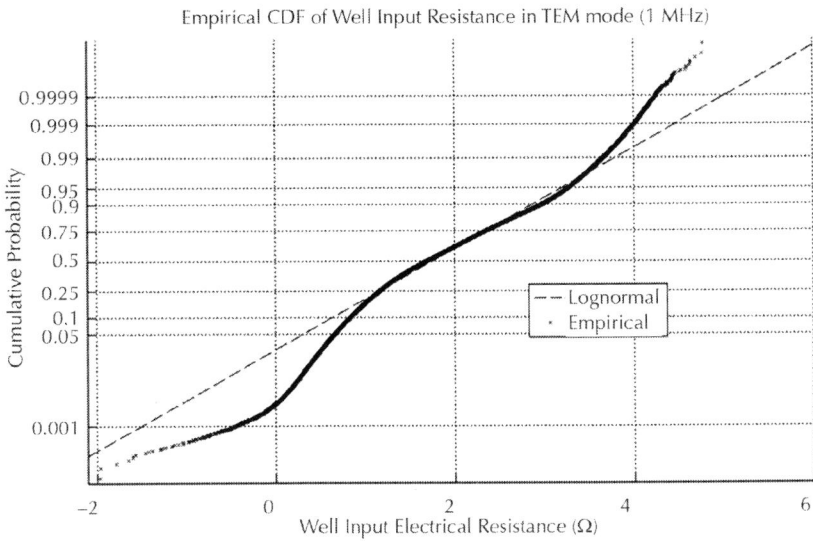

Figure 4: Empirical cumulative distribution function of the well input resistance in TEM mode at 1/4 wavelength from well head. The horizontal axis show the log on base 10 of the value of input resistance in Ohms, while the vertical axis show the cumulative distribution function value.

By performing the attenuation analysis, it was concluded that for 95% of the cases, the constant of attenuation in TEM mode is less than 0.8×10^{-4} Np/m for a frequency of 1 MHz. The power attenuation at 4000 m depth was also observed to be approximately 100 dB for the same frequency of 1 MHz, with 95% probability. Although 100 dB appears to be a large attenuation, most wells in operation are between 1000 m and 2000 m in length, and the attenuation is much lower at these depths.

REFERENCES

1. S. Brilles, "Remote Downhole Well Telemetry," US Patent US6766141, 2004.
2. J. A. D. Rosa, A. J. Carvalho and R. D. S. Xavier, "Petroleum Reservoir Engineering," Rio de Janeiro, 2006.

3. F. Sakata, H. Wakiwaka, M. Hanabusa, N. Yamazaki and H. Yamada, "Performance Analysis of Long Distance Transmitting of a Magnetic Signal in a Cylindrical Steel Rod," IEEE Translation Journal on Magnectics in Japan, Vol. 8, No. 2, 1993, pp. 102-106.
4. F. Harold J. Vinegar, R. R. Burnett, G. C. W. M. Savage and J. W. Hall, "Permanent Downhole, Wireless, TwoWay Telemetry Backbone Using Redundant Repeaters," US Patent US6633236B2, 2003.
5. B. W. Kennedy, "Energy Efficient Transformers," New York, 1998.
6. K. A. Safynia and R. W. McBride, "System and Method for Communicating Signals in a Cased Borehole with Tubing," US Patent US4839644, 1989.
7. P. A. Patil, et al., "Experimental Study of Electrical Properties of Oil-Based Mud in the Frequency Range from 1 to 100 MHz," SPE Drilling and Completion, Vol. 25, No. 3, 2010, pp. 380-390. doi:10.2118/118802-PA
8. C. A. Balanis, "Advanced Engineering Electromagnetics, Vol. 52, No. 1," Wiley, Hoboken, 1989, p. 1008.

Chapter 8

No-moving-part Valve for Automatic Flow Switching

Václav Tesař

Department of Thermodynamics, Institute of Thermomechanics v.v.i., Academy of Sciences of the Czech Republic, Dolejskova 5, 182 00 Prague, Czech Republic

ABSTRACT

Flow control valve containing no moving components was developed for switching fluid flow passing through it into a parallel secondary path once the conditions in the main path reach a certain limit. No sensors or actuators are involved; the switching is caused by the inability of the Coanda effect to keep the flow attached to a wall that leads it into a particular outlet once a large resistance is met in this outlet. So far, there has been no method for designing such valves. To provide a guidance, this paper summarises experimental evidence about loading characteristics obtained with a considerable number of tested valve geometries.

INTRODUCTION

A requirement encountered quite often in systems with flowing fluids is quick switching of the flow in response to some change of the downstream state of the system. Common electronic control with actuated valves as a rule is not as quick as the flow system. Especially in critical situations, the quick action is demanded combined with the valve being fail-safe.

Solution to this difficult problem was found by author in fluidic flow control valves operating without moving parts. They are very reliable and need no maintenance. Also, they may be easily built as extremely robust and resistant to adverse conditions. It is possible to manufacture them from refractory or other difficult-to-machine materials and if the material is chosen properly, the valves can operate at extreme temperature, vibration, or nuclear radiation levels. The switching dependent on the downstream conditions is obtained in an interesting – and yet little known – category of fluidic valves. This is a segment of a more general class of fluidic devices that perform automatically certain control tasks without using any of the traditional control circuits approach: there are no sensors, no actuators and similar control-loop components, in traditional approach seemingly inevitable. Typically operation of these passive control valves is based on their special shape of loading characteristic. This is e.g. the case of the pressure regulator valve described in [1]. Another example is the fluidic valve in Rolls-Royce aero engine limiting the efflux of compressed air through a failed pipe [2]. Yet another valve was described in [3] to work with fluid as extremely difficult to handle as are molten metals. Of course, the operation without an external control signal, using just the hydrodynamics of a particular internal geometry, makes such valves exceptionally resistant and suitable for the most severe operating conditions – e.g., for handling chemically aggressive liquids – a requirement often encountered in chemical engineering. The particular valve discussed in this paper was developed by the author to operate at high temperature levels – in fact glowing bright red. Its role, as shown in Fig. 1, is to protect the first-stage catalytic reactor (designed to operate instantly but only below certain temperature levels) by switching the flow of processed hot gas into a by-pass and this way into the second-stage reactor. The catalyst in the first stage has a relatively narrow window of temperatures in which it can operate efficiently. In particular, it is absolutely necessary to operate it so as to avoid the upper end of the window, beyond which the catalyst is in danger of irrecoverable changes. If the first stage is by-passed, the processing of the gas is done in the downstream second-stage reactor, with higher light-off temperature level and therefore unable to perform properly in those regimes where the first stage is active.

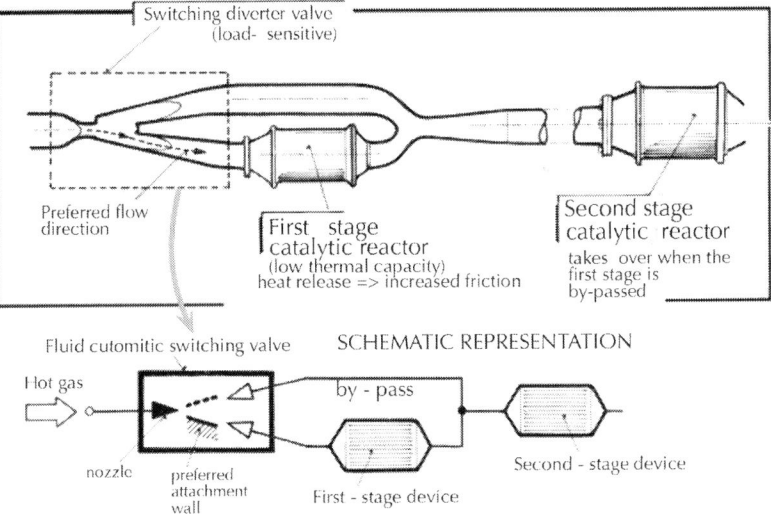

Figure 1: The discussed diverter valve was developed for switching hot gas flow into the by-pass as soon as it's increasing temperature – augmented by exothermic character of the reaction in the first-stage reactor – reaches the level at which the catalyst is in danger of an irreversible change.

The valve performing this flow-diverting operation is extremely simple. In principle, it consists of only two plain components: a nozzle forming a jet and an attachment wall to which the jet adheres due to the Coanda effect. The latter is sensitive to the loading of the valve output and ceases to keep the jet attached at a certain level of the hydraulic resistance met in the connected device(s).

In the typical example of an application presented in Fig. 1, the system was designed to operate starting from lower supplied gas temperature, gradually increased and rising also due to the heat generated in the reactor due to the exothermic character of the reaction. At the lower temperatures, the jet issuing from the nozzle is guided into the first-stage reactor. As the gas temperature increases, so does its viscosity and this leads to increased hydraulic resistance opposing the flow in the small cross-section passages on the surfaces of which is immobilised the catalyst. The resistance makes the action of the Coanda attachment more difficult as it is much easier for the jet to leave the valve through the by-pass, where it does not meet such opposition. The essential problem for designing this fluidic circuit is adjustment of properties of the valve so that the loading at its output puts the critical measure of strain on the Coanda effect at exactly the right temperature [4].

This may seem to be a rather exceptional case of operating conditions – but, in fact, a similar need for the automatic switching action at a certain level of gradually increased loading is actually encountered quite often, as is documented by the second example presented in Fig. 2. This is a filtering system – operating at essentially constant temperature. The problem which the valve solves is the (usual) clogging of the filter by the particles (such as, e.g., dust) removed from the fluid. This increases the hydraulic resistance the flow meets. When the Coanda effect finally fails in its directing the flow into the main (primary) filter, the fluid flow is switched into the clean back-up filter – again automatically, without any external monitoring sensors and actuators.

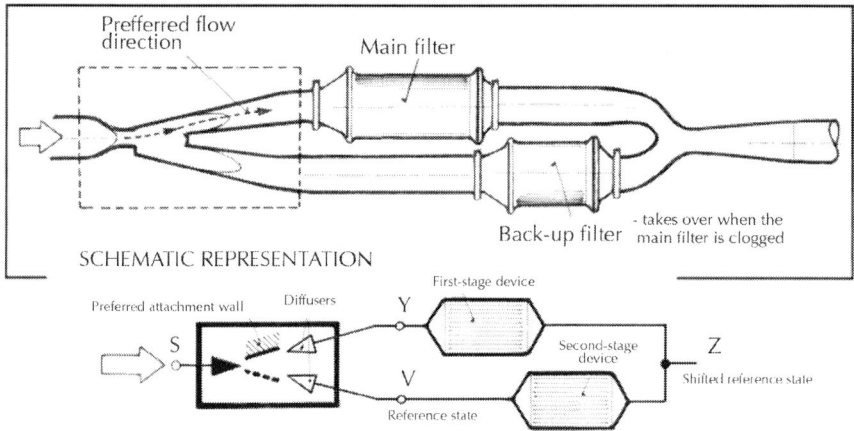

Figure 2: Another example of the use of the automatically switching diverter valve, here removing the danger of the flow decreasing below an acceptable minimum due to the clogging of the main filter. With the back-up (the second stage) filter in place of the empty by-pass of the previous case (Fig. 1), the circuit design is slightly more complicated.

The need to accelerate the fluid flow in the nozzle to generate strong enough Coanda effect may lead to high energetic loss in the valve (the loss being typically proportional to the square of flow velocity) and it is generally advisable to provide inside the valve some means for converting the fluid kinetic energy back into at least a part of the pressure energy the fluid had before entering the nozzle. This is the reason why there are in the valve the diffusers; their requirement of small divergence angle makes them quite long and at first glance the prominent parts of the valve. Apart from the main, preferential attachment wall there may be another wall on the other side of the jet path, guiding the flow into the vent terminal after the flow is switched.

The valves with automatic diverter switching in dependence on the loading are no new invention; they have been known and used – though only very sporadically – for more than two decades. Apart from the fact that the idea is little known, the reason for their rare use – much less than their undeniable advantages would suggest – is the lack of a design method sufficiently simple or at least straightforward. The loading properties upon which their operation is based, and the dependence of these properties on the geometry of the internal space (with its innumerable choices of various dimensions) are far from being understood. In all applications so far, the particular design has been a result of painstaking cut-and-try process. This paper aims at elucidating at least some of the relations between some of the design parameters in the resultant valve behaviour.

LOADING CHARACTERISTIC OF THE VALVE

In contrast to mechanical flow diverting valves with moving parts, in which the fluid is prevented from entering a particular flow path under any conditions – by blocking materially the entrance into it – operation of fluidic flow control diverter valves without moving parts depends on more subtle balance. Proper conditions in the fluidic circuit have to be carefully set up, taking into the consideration the properties of the valve as well as those of the device connected as the load to its output terminal. The adjustments are best made by following the mental guidance of a graphical representation. One of the reasons why this is so useful is the usual fact of the properties of fluidic devices being typically strongly non-linear and therefore not easily expressed by a simple formula.

For the purposes of the circuit design the valve properties are most usefully described by graphical presentation of the loading characteristics – a dependence of the output specific-energy difference on the output flow rate – a generic example of which is presented in Fig. 3. In contrast to a mechanical valve, typically delivering to its output terminal exactly all the fluid flow supplied into the inlet, the jet-type fluidic valve properties are not so simple. Their output flow rate in the output terminal Y may be higher than the supplied flow rate into S since the jet generates the jet-pumping effect, entraining additional fluid through the vent terminal V. On the other hand, throttling the output flow – which may be done in an attempt to obtain higher specific energy in Y (and therefore higher available acting pressure, since the pressure energy component usually represents most of the available energy of the fluid) – causes spilling of some supplied fluid over into V. The end of the characteristic curve approached by the throttling is of particular importance for the subject of this paper: it represents the state in which the gradually increased throttling (by decreasing available cross-section in the connected load) overcomes the capability of the Coanda effect to keep the flow attached to preferred attachment wall and all the flow is switched into the vent V.

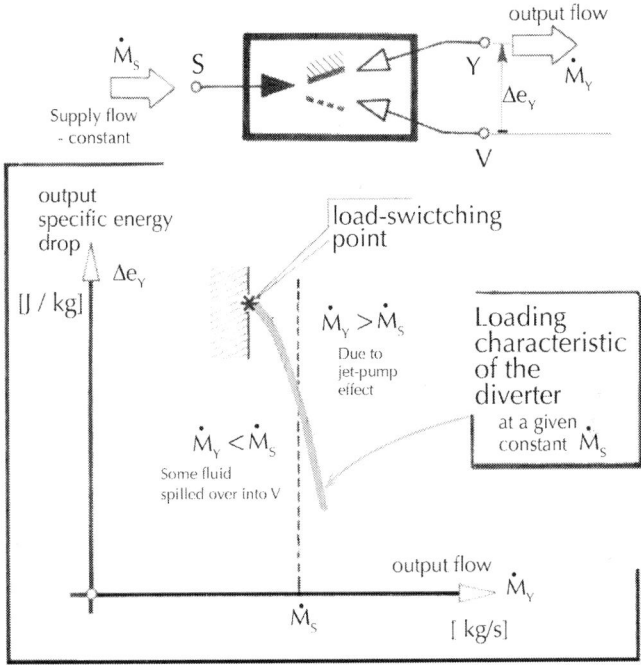

Figure 3: The loading characteristic represents the decrease of the available output specific energy with increasing output flow rate. From the point of view of the subject of this paper, most interesting is the load-switching end of the characteristic.

Fig. 4 presents graphically the conditions in the circuit from Fig. 1 (the case with empty by-pass; for simplicity of the presentation the hydraulic loss in the by-pass pipe is here neglected). In this case the output specific energetic difference between the valve terminals Y and V is equal to the specific-energy drop e across the load. This is dependent on the flow rate passing through the load; the dependence is the *characteristic* of the load. In fluidics in general, especially at high Reynolds numbers, the characteristics are usually more or less quadratic, capable of being fully described by the value of the quadratic dissipance Q:

$$\Delta e = Q\dot{M}^2. \tag{1}$$

In the present case the characteristics of the load devices – those which are strongly influenced by viscosity changes in long narrow channels – tend to require also a considerable linear term [5] and [8]. The coefficient in this term may be expressed as a multiple of Q:

$$\Delta e = Q\dot{M}^2 + rQ\dot{M}. \tag{2}$$

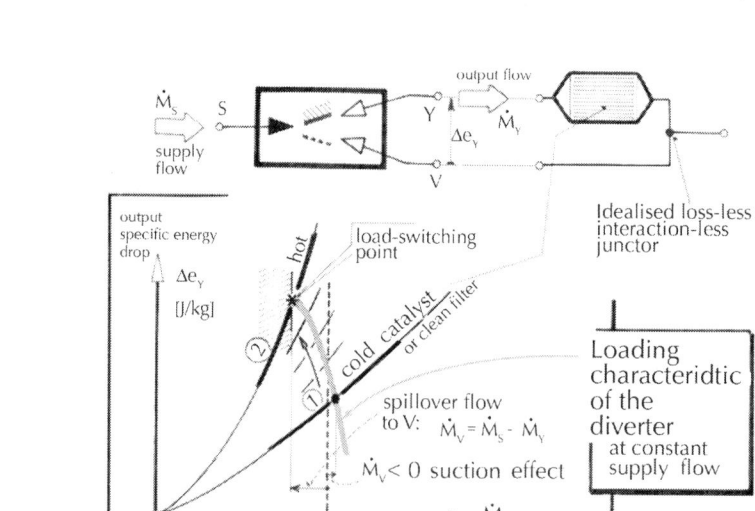

Figure 4: The changes in the circuit from Fig. 1, evaluated graphically as the successive positions of the intersection of the valve loading curve (Fig. 3) with the characteristic of the load, gradually varying from (1) to (2).

The gradually increasing e at the same flow rate due to clogging or viscosity increase is represented in Fig. 4 by the succession of the device characteristics progressing from (1) to (2) (those between these extreme cases are shown only in part). The states in the circuit are determined by the intersections with the characteristic of the valve. In the initial cold-start state (small resistance) there is in the case represented in Fig. 4 a small negative flow (i.e. directed backwards into the valve) in the vent terminal V (through the by-pass). This means the flow passing through the loading device is larger than the flow rate supplied into the nozzle. This effect may be avoided by a different choice of properties of the devices. Alternatively, this re-circulation of the fluid in the valve exit loop may be useful: e.g., in the case of a filter this ensures removal of any particles that manage to get through during the first pass – and in the case of a catalytic reaction the re-circulation increases the yield. Gradually, the conditions usually change into the regime of considerable spillover into the vent and finally approach the load-switching state.

The characteristic of the valve (as presented in Fig. 3) is, of course, useful also in the case with a loading device placed into each of the two outlet flow paths, according to Fig. 2. The graphical presentation used to evaluate the load-switching conditions for this case is shown in Fig. 5. The characteristic of the main, first-stage device positioned in the preferred exit Y is shifted down by the distance evaluated according to Eq. (2) for the parameters Q_2 and r_2 of the second-stage device as well as for the spillover flow $\dot{M}_S - \dot{M}_Y$.

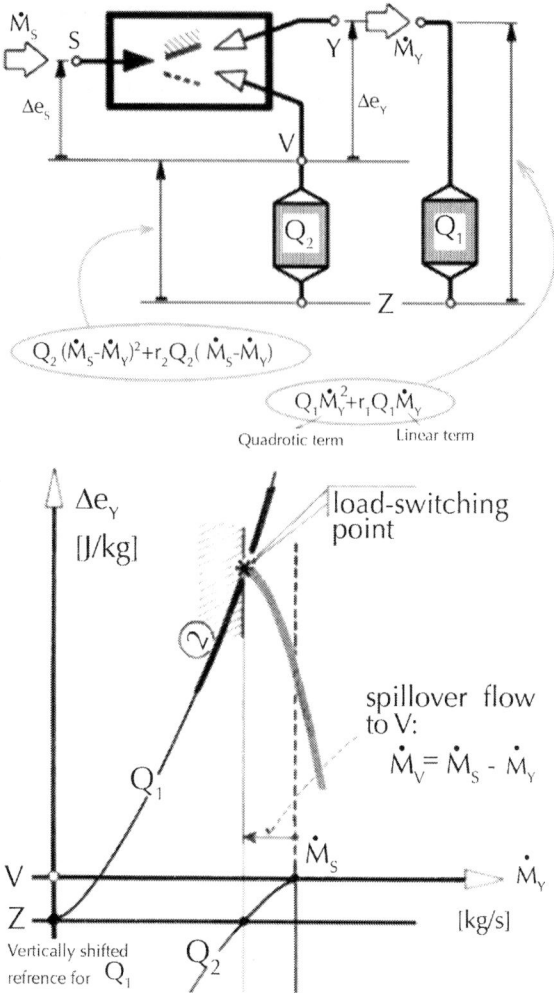

Figure 5: In the case of Fig. 2 the load-switching conditions are evaluated from the intersection of the valve loading curve with the characteristic of the load in its hot (or clogged) high resistance state (2) shifted down by an amount representing the specific-energy drop across the second-stage device Q_2.

Calculations follow the graphical representations – in the cases discussed above they are performed with absolute values of the flow rate and fluid specific energy. For general description of the valve loading properties it is useful to transform the loading characteristic into the dimensionless co-ordinates as they are presented in Fig. 6. Because of the Eulerian similarity [10], the whole family (theoretically consisting of infinitely many members) squeezes into a single universal curve [11] – perhaps with some small non-coincidences due to Reynolds-number dependence. The output flow rate is non-dimensionalised by relating it to the supply flow rate

$$\mu_Y = \frac{\dot{M}_Y}{\dot{M}_S} \tag{3}$$

And, in a similar manner, the output specific-energy difference is non-dimensionalised by being related to the supply specific-energy difference Δe_S (between the supply terminal S and the reference V)

$$\eta_Y = \frac{\Delta e_Y}{\Delta e_S}. \tag{4}$$

A fact not universally recognised is internal flowfield in the valve changing with the loading. This means also the supply specific-energy difference (Fig. 5) Δe_S varies as the relative flow rate μ_Y is changed. As a result, the shape of the resultant dimensionless loading curve

$$\eta_Y = f(\mu_Y) \tag{5}$$

Differs from the shapes of the original loading curves in absolute co-ordinates

$$\Delta e_Y = f(\dot{M}_Y). \tag{6}$$

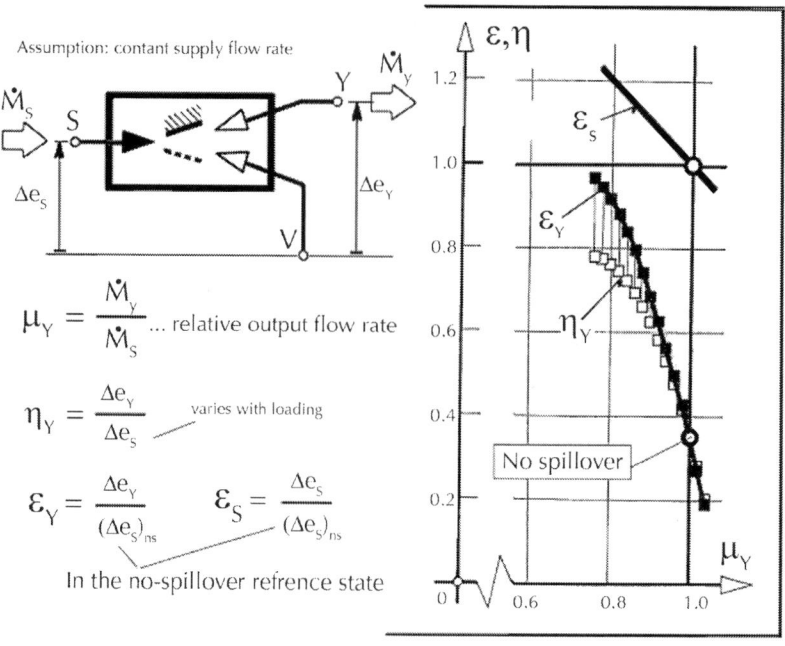

Figure 6: Transformed dimensionless co-ordinates bring the advantage of universality – apart from small dependence on Reynolds number, the loading characteristic is practically the same for any magnitude of the supply flow rate.

This was first noted by Tippetts and Royle [9] (actually for the related bistable amplifiers), who recommended avoiding this by taking as the reference not the supply specific-energy difference in the particular evaluated state but, instead, the supply specific energy $(e_s)_{ns}$ in the no-spillover regime, i.e. at the relative output flow $\mu_Y = 1$. This means evaluating a different dimensionless output specific energy:

$$\varepsilon_Y = \frac{\Delta e_Y}{(\Delta e_S)_{ns}}. \tag{7}$$

Also, it should be noted that it is possible (and useful) to evaluate the analogous expression for the variations of the relative difference between the supply terminal S and V

$$\varepsilon_S = \frac{\Delta e_S}{(\Delta e_S)_{ns}} \qquad (8)$$

The dependence of which on the relative flow rate

$$\varepsilon_S = f(\mu_Y) \qquad (9)$$

Should be also investigated and plotted in the full dimensionless output characteristic diagram.

The example at the right-hand side of Fig. 6 shows the difference in the curve shapes in the two dimensionless presentations by Eq. (5) and by

$$\varepsilon_Y = f(\mu_Y) \qquad (10)$$

For a simple linear variations of Δe_s with the loading. In the following discussion of experimental results obtained with different tested valve variants, the attention is focused on the shape-retaining non-dimensionalisation with respect to $(\Delta e_s)_{ns}$, i.e. using Eqs. (9) and (10). Because of the Eulerian similarity being not perfect, each such data set should be accompanied by a statement about the value of the test Reynolds number, usually evaluated for the conditions in the exit of the nozzle.

TESTED MODEL VALVE

The valve used in the laboratory investigations was derived from a design intended for operation at a rather large scale, processing exhaust gas flow [7] in a system of devices [5] connected by 2.5 in. outer diameter pipes. This size was retained in the model, but the tests did not simulate the exhaust gas system conditions and were performed with steady, non-pulsating cold air flow. It should be noted that pulsations, such as those encountered in an internal combustion engine exhaust [7], would actually considerably change the results, since even at large divergence angles of the attachment walls the pulsating flow would tend to attach to both of them – a strongly pulsating flow would certainly not produce the desirable Coanda effect in the 20.6° wall divergence cavity of the present model shown in Fig. 7.

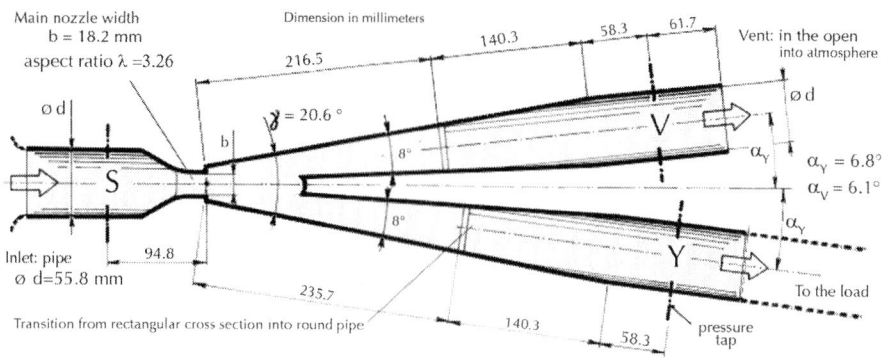

Figure 7: Overall geometry and dimensions of the full-scale laboratory test model, with round-pipe inlet as well as outlets while the central core part (of critical influence on the properties because the velocities there were high and aerodynamic effects most pronounced) was of rectangular, constant-depth cross sections.

The valve geometry was complicated by the transitions between the circular cross-sections in the terminals and the constant-depth geometry, i.e. rectangular cross-sections in the core part (Fig. 8). Due to the constant depth in this part, the geometry there is essentially two-dimensional. This, however, cannot be said about the internal flowfield there. The small, 20.6° total divergence was chosen in an attempt to minimise the hydraulic losses associated with change of flow direction.

To secure the property of the flow in unloaded valve always passing into the output terminal Y, the design as seen in Fig. 7 and Fig. 8 is visibly asymmetric. The inclination of the attachment wall is smaller on the Y side $a_Y < a_V$ (Fig. 7). Also there is a larger setback $c_Y < c_V$ (Fig. 8) on the vent side. The jet in the starting flow regime always attaches to the wall with smaller setback and smaller inclination angle – and this ensures that fluid is directed into the output terminal Y. Apart from this Coanda effect preference, in the basic design shown in Fig. 7 and Fig. 8 stabilisation of the deflected jet flow into Y, was augmented by providing the internal positive feedback loop – by the bi-cuspid nose of the splitter with the concave wall between the cusps, in a similar manner as was found useful earlier (e.g. in [12] and [13]).

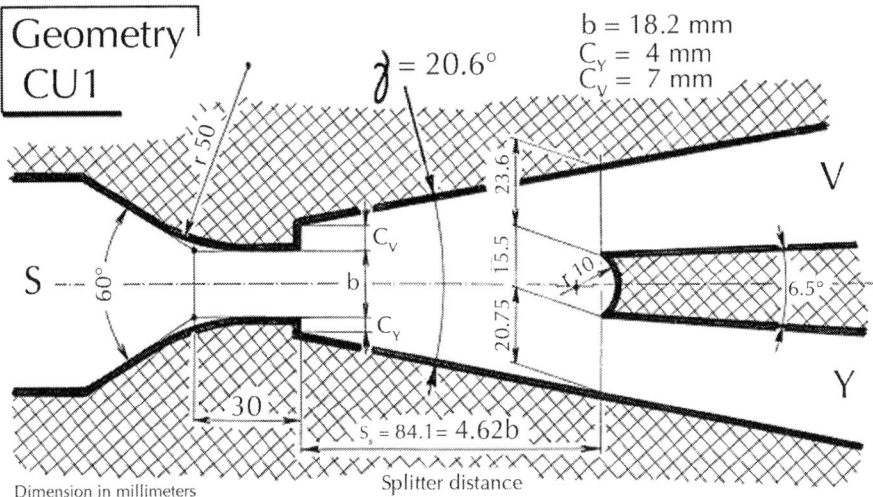

Figure 8: Geometry of the valve core in the first of the tested variants. The depth – perpendicular to the drawing plane – was $h = 52.8$ mm, the same everywhere throughout this core.

Present paper is devoted exclusively to the results obtained by experimental investigations. In fact they were supported by independent numerical flow field computations, performed using the standard FLUENT software in the fully three-dimensional computational domain, the bottom half of which is shown in Fig. 9. The fully 3D solutions were chosen because of the expected complications associated with the circular/rectangular/circular transitions in the flow path cross-sections. Considering the essentially two-dimensional geometry of the critical part of the valve as shown in Fig. 8 (and also considering the relatively large aspect ratio, making less influential the effects of the friction on the top and bottom flat walls there), no significant three-dimensionality of the flow field in the valve core was initially expected. The more surprising was therefore the fact of there being actually a very strong three-dimensionality, as documented in Fig. 10. Obviously, the computations have shown the pathlines in the collector leading to the vent V to be actually helical in shape. This explains the common failures of some fluidic device designers who attempt evaluation of the internal conditions by simple one-dimensional calculations based only on the size of the cross-sectional areas. In particular, the behaviour of the diffusers may be in some regimes significantly influenced by the complex character of the flow field at their entrances. The helical character of the flow inside the valve is an important feature that underlines the general importance of helical flow mode instabilities in fluid mechanics, as discussed, e.g., in [16] and [17].

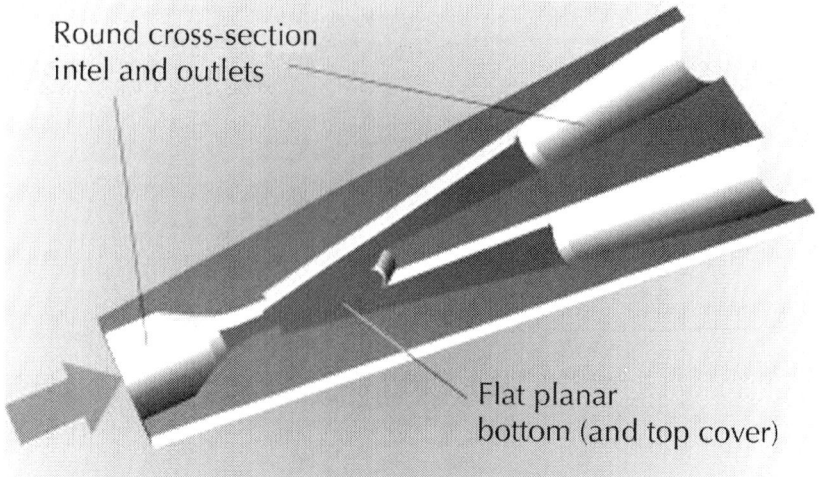

Figure 9: The lower half of the computational domain used in the numerical flowfield computations. The picture presents instructively the transitions between the round cross-sections of the inlet and outlet pipes and the rectangular (constant-depth) cross-section shape of the core.

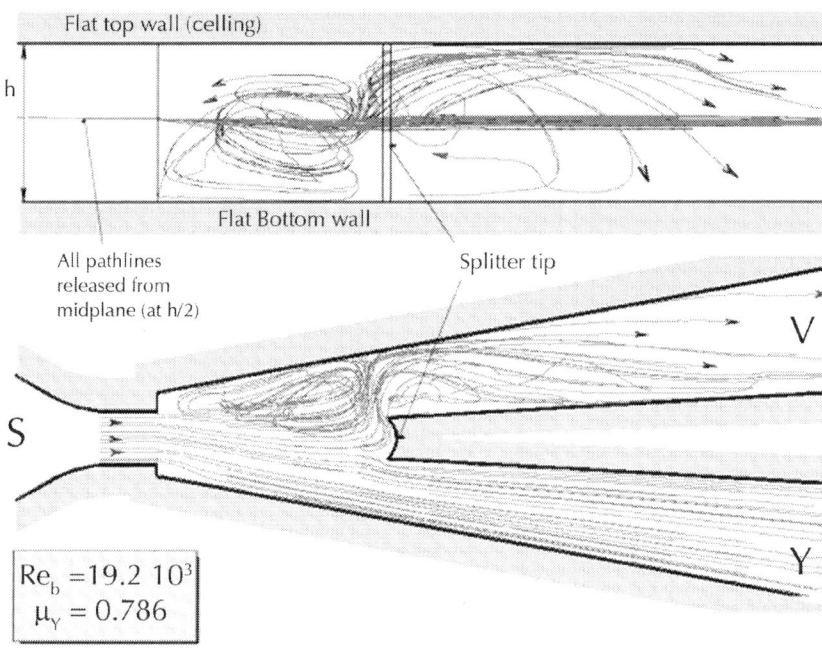

Figure 10: An example of computed pathlines obtained for rather typical conditions demonstrates the unexpected strong three-dimensionality of the flowfield. The pathlines (trajectories of imagined particles left to follow the local flow directions) were all released from the same midplane height and yet some of them soon reached as low as to the bottom and others as high as the top cover plate.

EXPERIMENTAL RESULTS WITH THE BASELINE GEOMETRY

The experiments with the valve in its initial geometry, as shown in Fig. 7, Fig. 8 and Fig. 9, were run on four occasions, at four different constant air mass flow rates supplied into the nozzle. The flow rates into the output Y were gradually more and more restricted and – together with the supply flow rate, measured and kept constant – were measured by an orifice flow meter. The pressure differences measured between the locations S, Y, and V (the positions of these locations are indicated in Fig. 7) were converted into the differences in specific pressure energy, to which were added the local values of the air specific kinetic energy computed for the same locations. Because of the relatively large cross-sections there, the pressure component was dominant in Δe_Y.

The values Δe_Y for the three test runs are plotted in Fig. 11 as a function of the mass flow rates. All three curves are mutually similar. In the following Fig. 12, the collected experimental data (including those from Fig. 11) are plotted converted into the dimensionless values using the expressions listed in Fig. 6. Indeed, due to the similarity, all the data in the transformed co-ordinates $\square_s = f(\mu_Y)$ fall on a single curve. Also plotted is the supply characteristic $\square_Y = f(\mu_Y)$ making the diagram Fig. 12 the complete output characteristic of the valve. A remarkable fact seen there is the substantial change in the character of the flow field, markedly revealed by the sudden kink in the supply curve during transition, in the no-spillover state $\mu_Y = 1$, from the regime with positive spillover into the jet-pumping regime. Also of interest in Fig. 12 is the position of the load-switching point and the hyperbolic line of constant relative output power, evaluated for the state just before switching.

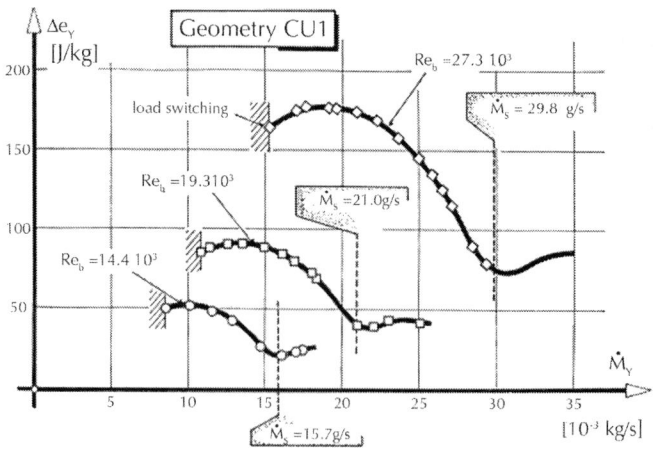

Figure 11: The loading characteristics in absolute co-ordinates according to Eq. (6) obtained in the experiments with the baseline model (Fig.

7 and Fig. 8). Three independent measurement runs were made, each at different constant supply mass flow rate (note the corresponding values of Reynolds number, computed from the nozzle exit width).

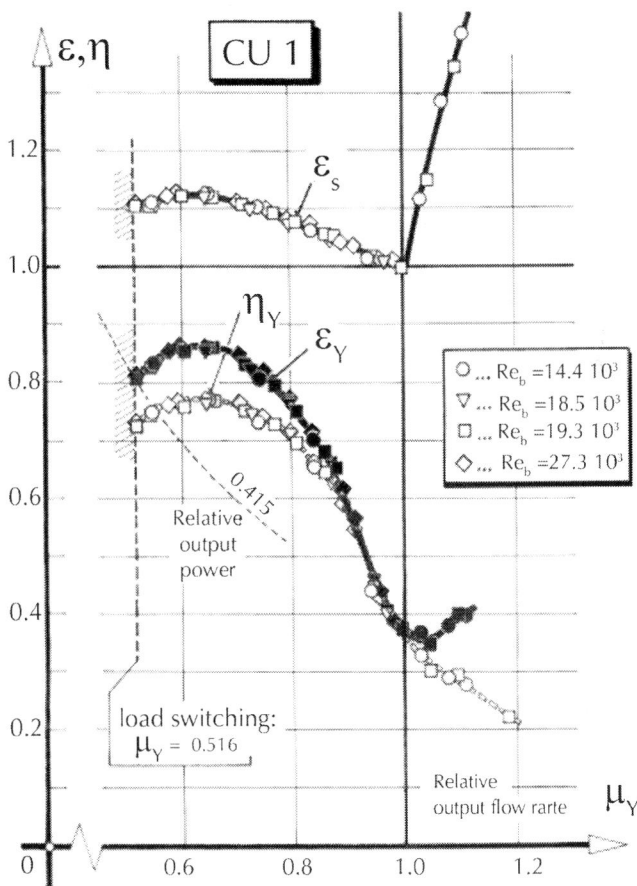

Figure 12: Demonstration of the validity of the Eulerian similarity: all experimental data obtained for the baseline valve (Fig. 7 and Fig. 8) fall on the same curves in the dimensionless co-ordinates Fig. 6.

Somewhat disappointing may be the relatively low efficiency of the valve – mere 37% in the no-spillover state. Obviously, because of the dominance of the pressure component in the specific-energy differences, this reflects a rather poor pressure recovery in the valve. This is rather surprising, considering the very short nozzle-to-splitter distance s_s = 4.62b (Fig. 8) much shorter than the usual values in fluidic amplifiers with much longer distance s_s and therefore providing more opportunity for the jet to lose its momentum. The traditional line of thinking is the pressure recovery to be mainly a matter of the conversion that takes place in the diffuser(s) and here the jet is guided into them without being given much opportunity to lose its momentum by mixing with slow external air. Admittedly, their relatively large 8° divergence angle (Fig. 7) makes the diffusers in this model not the most efficient ones, but the value is not so large as to make the diffuser performance really poor. Recent discovery of the dominant role of a different pressure recovery mechanism [13] was, of course, not known at the time when the valve was designed. It is probable that it is an unfavourable condition for this trapped-vortex mechanism that was the real reason for the disappointment.

Obvious pre-requisite for the remarkably good similarity demonstrated in Fig. 12 is the small dependence of the character of the internal flow field on Reynolds number. This dependence, experimentally found for the valve from Fig. 7 and Fig. 8, is presented in terms of Euler numbers and valve efficiency in Fig. 13 – as functions of Reynolds number evaluated from the nozzle exit conditions. The values plotted there were all obtained in the no-spillover regime. Indeed, most of the values there are very near to a constant, at least at the sufficiently high Reynolds numbers as were those indicated in Fig. 12 as the conditions of the experimental runs. Significant deviations from the quadratic similarity were found only at very low Reynolds numbers. Especially remarkable there is the increasing efficiency at low Re, a feature that may be of some significance for the low Reynolds numbers microfluidics (e.g., [14]).

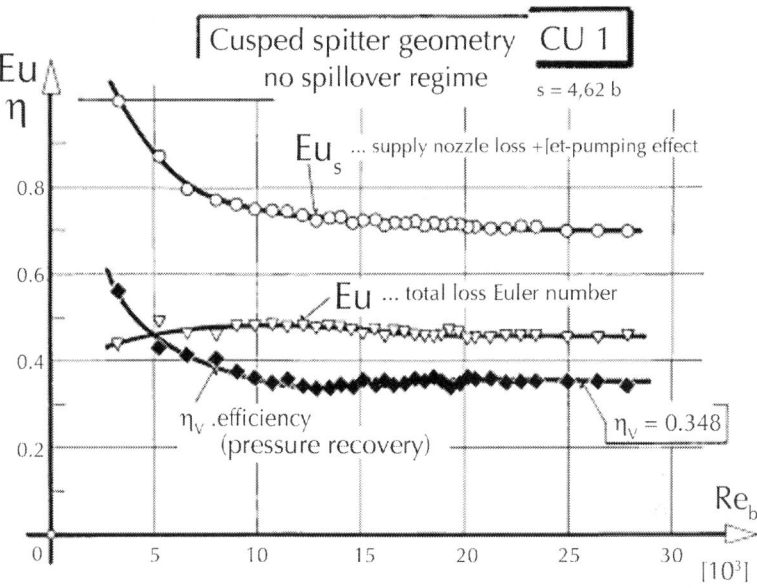

Figure 13: Experimental dependence of the loss coefficient (Euler numbers) and valve efficiency of the baseline CU1 model on Reynolds number.

CHARACTERISATION BY THE THIRD-ORDER TENSOR

The graphical representation of properties by the characteristic curves, as shown in Fig. 12, provides an instructive guidance for the circuit design, but is not suitable for working with it on a computer. For the purposes of storing the data and performing the design calculations it is preferable to have an analytical description of the device.

In electronic circuits, the properties of analogous three-terminal devices for such purpose are usually expressed by a matrix that defines a linear dependence between the vectors of currents and voltages. This is not applicable in the non-linear case of the fluidic valves. What may be taken over is only the essential idea of the characterisation by the dependence between the vector of mass flow rates and a corresponding vector of specific-energy drops (Fig. 14). Because of the usually dominantly quadratic character of the dependences in fluidics, the dependence is with sufficient accuracy expressed using the vector-to-vector quadratic dependence [15]

$$\Delta e_i = \sum_j \sum_k Q_{ijk} \dot{M}_k \dot{M}_j \tag{11}$$

As shown in Fig. 14. The third-order tensor that defines this dependence is the characterisation tensor Q_{ijk} containing the multiplicative coefficients. In fact, the individual terms of these coefficients may be easily demonstrated to possess reasonable physical meanings, a listed in Fig. 15. The terms Q_i characterise, similar as does the scalar value of quadratic dissipance Q in Eq. (1), the energetic losses in the individual parts (components) of which the valve consists: the nozzle and the two diffusers. Apart from the loss terms, there is also the suction term S. If it were absent, the tensor Q_{ijk} consisting of only the quadratic dissipances would be symmetric. It is the anti-symmetric part (the "suction" – actually here involving all the more complex effects of jet flow interactions) that makes this tensor approach suitable for the description of the valve as a bifurcator element. It is possible to extract the most important and easiest to evaluate dissipance – the value Q_1 for the nozzle – so that for the description of all characteristics in the relative co-ordinates it is necessary to evaluate just the three numerical values of the relative magnitudes of the terms:

$$q_2 = \frac{Q_2}{Q_1} \tag{12}$$

$$q_3 = \frac{Q_3}{Q_1} \tag{13}$$

And

$$s = \frac{S}{Q_1}. \tag{14}$$

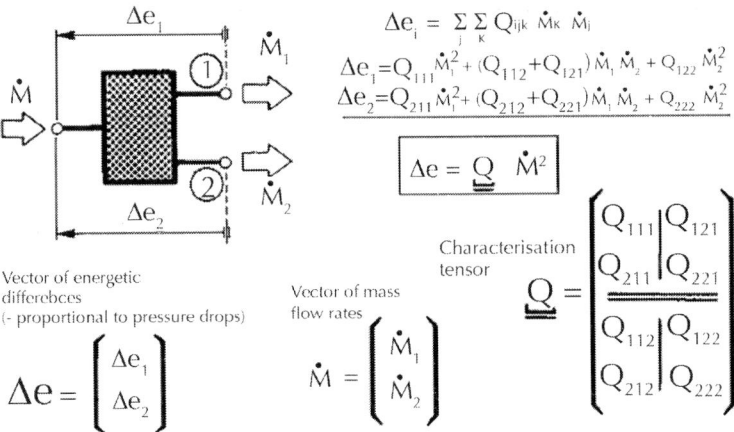

Figure 14: A general concept of bifurcator element (with one inlet terminal and two outlets) [18]. Its dependence of the vector of energetic drops on the vector of mass flow rates is defined by the third-order multiplication tensor of general quadratic dependence between two vectors.

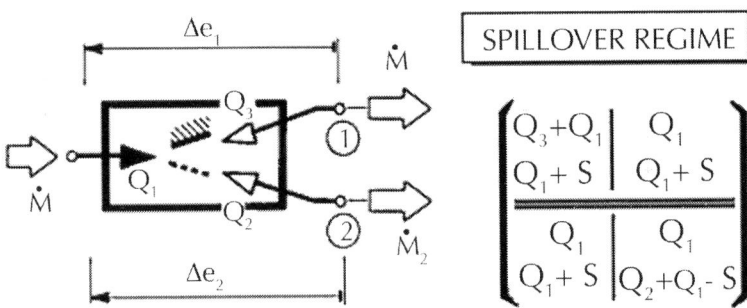

Components of the characterisation tensor:

Q_1 ...Quandratic dissipance of the nozzle

Q_2 ...Quandratic dissipance of the first collector

Q_3 ...Quandratic dissipance of the second collector

S ...Suction (jetpumping) effect

Figure 15: The valve in the spillover regime is a case of the bifurcator (Fig. 14). The components of the characterisation vector in this case possess a simple physical meaning.

Their magnitudes were identified for the valve in its baseline geometry CU1 and are presented in the upper-right corner of Fig. 16, the rest of this picture showing how successful is this description of the valve properties compared with the graphical presentation of the loading characteristic.

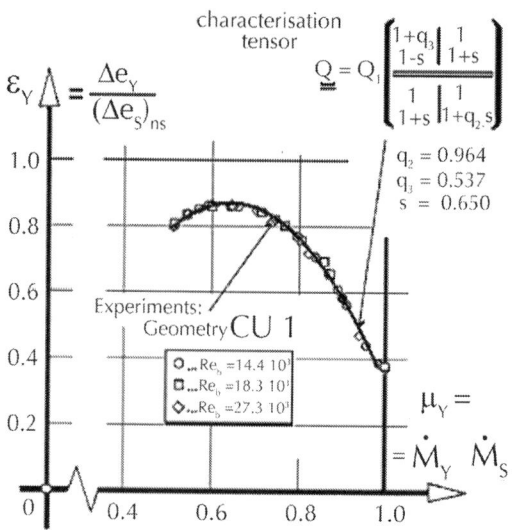

Figure 16: It is possible to identify the values in the third-order characterisation tensor of Fig. 15 the insertion of which into the general quadratic vector-to-vector mapping function of Fig. 14 produces a loading characteristic exactly corresponding to the experimental findings.

CHANGING THE SPLITTER DISTANCE

The jet leaving the nozzle in the valve loses its momentum – and specific kinetic energy – by mixing with the slower outer fluid in the interaction cavity. Compared to this, the loss due to its surface friction when it is attached to the inclined attachment wall, though considered by some inexperienced observers a source of hydraulic losses in the valve, is relatively insignificant. There is a general consensus that to keep the valve efficiency at a reasonable level, the jet flow has to be converted as soon as possible into the wall-bounded flow in the diffuser. For this, the valve geometry should be designed with as short as possible nozzle-to-splitter distance ss. Some distance travelled by the jet, nevertheless, is necessary to provide it with an opportunity to the switching motions. In the successful bistable diverter valve reported in [12] the relative value of this distance was

$$\sigma = \frac{s_s}{b} = 6.85 \qquad (15)$$

And this was considered exceptionally small value compared with other known Coanda effect valves. In Fig. 8, this relative value for the geometry CU1 is shown to be even much less, $\sigma = 4.62$, and yet the efficiency was found to be low, mere $\eta = 0.348$ as the asymptotic value in the no-spillover state (Fig. 13).

It seemed rather obvious to investigate what improvement, if any, was achievable by making the splitter nose even longer. The tested geometry CU2 with the lengthened splitter resulting in the value σ as small as $\sigma = 3.15$, is presented in Fig. 17. Its properties, as shown in Fig. 18 and Fig. 19, were a disappointment, its asymptotic efficiency value only $\eta = 0.203$.

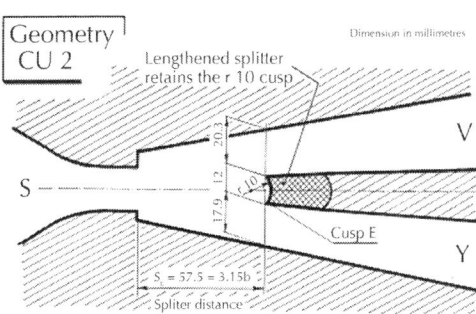

Figure 17: Geometry of the baseline model adapted to very short distance s_s from the nozzle exit to the splitter nose. According to general belief, the shorter length of jet losing momentum by mixing with slower outer fluid should result in higher efficiency.

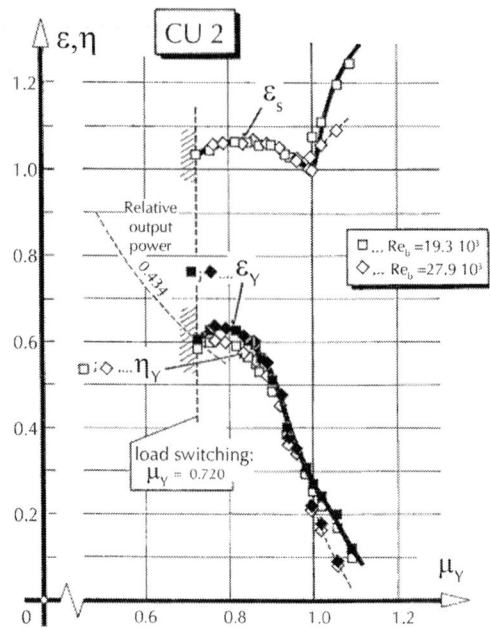

Figure 18: Experimental complete output characteristic, in similarity co-ordinates, of the valve in the geometry variant CU2 as shown in Fig. 17.

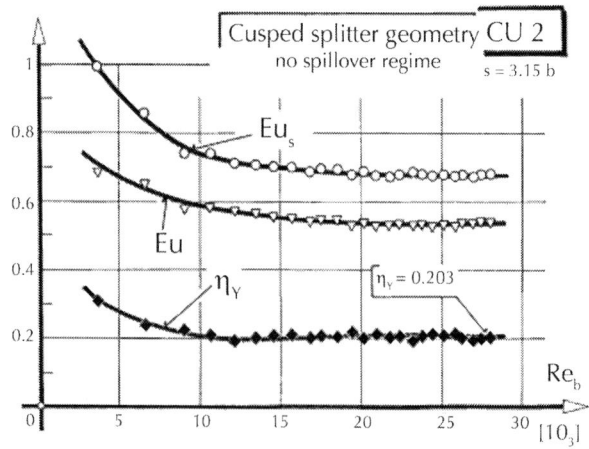

Figure 19: Experimental Reynolds-number dependence of the loss coefficient (Euler numbers) and valve efficiency for the variant CU2 from Fig.

17. Comparison with Fig. 13 indicates a larger proportion of Reynolds-number dependent friction loss component and – contrary to expectation – lower efficiency.

The values of the coefficient Eqs. (12), (13) and (14) were also evaluated and are presented in Fig. 20 in comparison with the original baseline geometry CU1. What the lengthened splitter results in is just an increased relative dissipance q_2 of the first (preferred) collector. Since the general character of the splitter nose shape is similar, it is in this case possible – as shown in Fig. 20 – to guess rather safely the probable properties of the four interpolated cases. Extrapolation is, of course, less safe but on the basis of what is presented in Fig. 20 it may be suggested that a considerable efficiency improvement might be in this case – contrary to common expectations – obtained by choosing the relative distance larger. This seems to be an obvious consequence of the dominance of the captive vortex pressure recovery mechanism as discussed in Ref. [13].

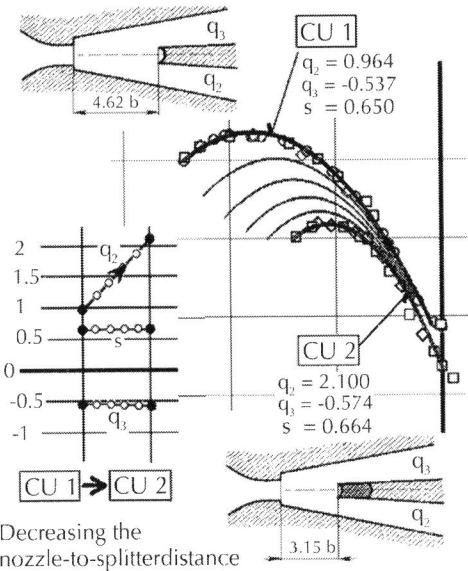

Figure 20: Comparison of the loading curves for the two cusped-splitter geometries. Identified numerical values of the coefficients in the characterisation tensor of Fig. 15 show the lengthened splitter results just in in-

creased dissipance of the first (preferred) collector – and the corresponding deterioration of valve properties.

ROUNDED SPLITTER NOSE

A characteristic feature of jet-deflection fluidic devices designed by the present author [6], [12] and [13] is the bi-cuspid nose of the splitter between the two outlet passages. The wall connecting the two cusps is of concave shape. If the jet is deflected by the Coanda effect towards the preferred attachment wall, its outer layer – on the opposite side to the wall – hits one of the cusps, the cusp E shown in Fig. 17. The outer layer is peeled off and turned by following the concave part of the splitter nose, which it leaves in a direction almost opposite to the flow direction of the main part of the deflected jet. The back-turned part flows towards the other, unused attachment wall (the one directed into the vent V) which turns it even more until its trajectory forms nearly a full circle. In the last part of its return trajectory, this backwards flowing fluid is attracted towards the initial part of the main jet due to its entraining, jet-pumping influence.

It may also hit upon the main jet by whatever remains from its initial momentum. As a result, the interaction cavity of the valve downstream from the nozzle exit is filled with a standing, captive vortex. This is well visible in the bottom part of Fig. 10. The vortex may entrain some fluid from the vent at $\mu_Y < 1$ or (as in Fig. 10) at $\mu_Y < 1$ may be bleeding some fluid as the spillover flow into V. Nevertheless, over most of the valve operating range (before the switching occurs) it stays there as an important entity influencing the internal flow field. In principle, it acts as a positive feedback loop, preventing the jet from separation from the preferential attachment wall. If the jet starts to separate from the attachment wall, its trajectory will be more straight, leading towards the cusp E. The outer layer peeled off by this cusp is more powerful and on its return to the initial part of the jet pushes it more strongly back towards its attachment wall.

Of course, this mechanism extracts more power from the jet than if it were not required to flow past the sharp edge. It may seem to be a reasonable guess that with a smooth, rounded splitter the efficiency would be higher. The opportunity to test this idea was used and an alternative, round-nose splitter geometry RD1 was extemporised as shown in Fig. 21. Without the positive feedback loop, the jet is expected to separate from the attachment wall and load-switch more easily but, after all, the jet attachment in the baseline geometry CU1 is seen in Fig. 12 to be overly stable – the switching there is not a very effective process if it takes place in a situation where already 48.4% of the flow spills over into the vent terminal.

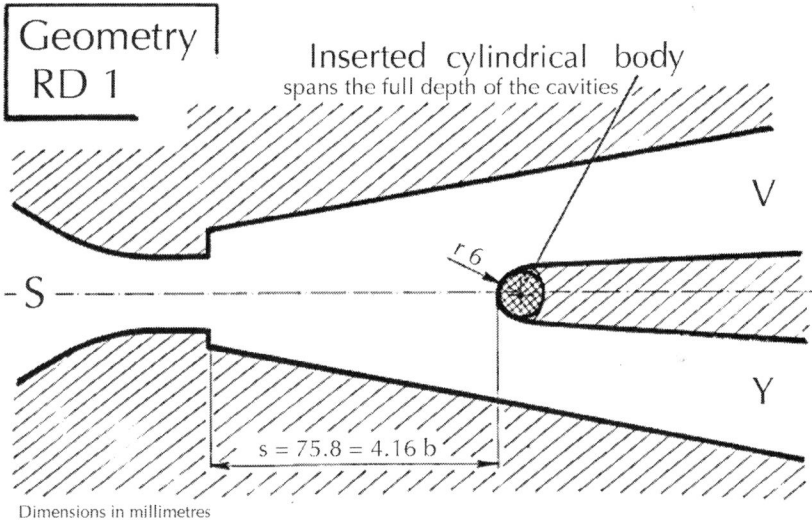

Figure 21: Valve geometry adapted to convex splitter shape intended to answer repeated questioning of the wisdom of author's bi-cuspid splitter nose [12].

With the rounded splitter RD1, the output characteristic presented in Fig. 22 shows switching indeed occurs at a smaller spillover flow. Comparison of the following Fig. 23 with Fig. 13 shows even some efficiency improvement (44.5% against 34.8%)

in the reference state $\mu_Y = 1$, but this is a rather exceptional situation and the overall comparison in Fig. 24 presents a rather unfavourable picture.

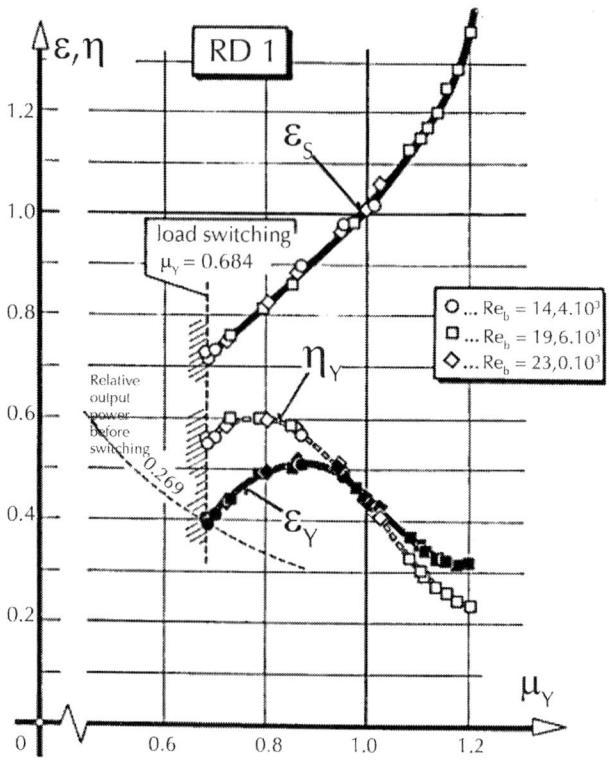

Figure 22: Experimental complete output characteristic, in similarity coordinates, of the valve in the adapted version with rounded splitter geometry RD1.

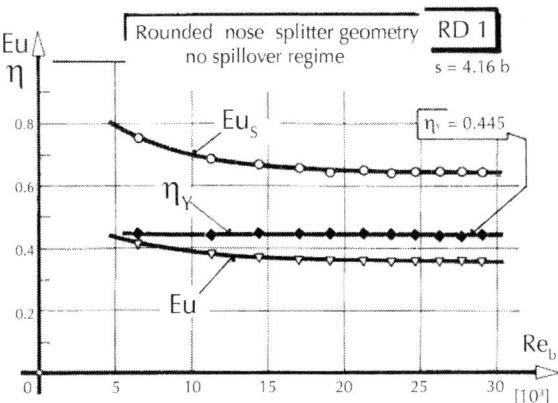

Figure 23: Experimentally determined loss Euler numbers and efficiency for the valve variant shown in Fig. 21. In comparison with Fig. 13, the rounding of the splitter brings some improvement in efficiency which, however, loses its interesting trend of improving with decreased Reynolds number.

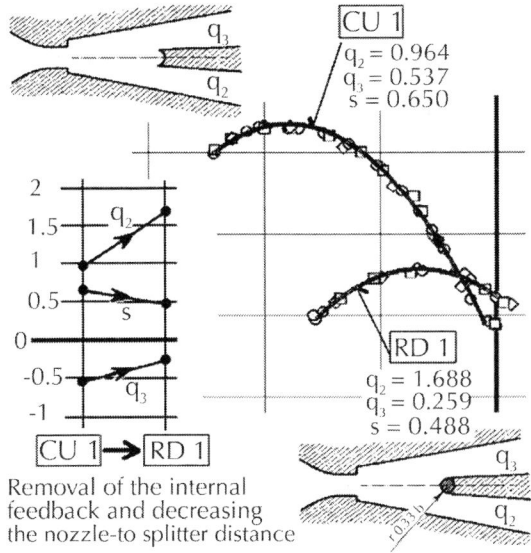

Figure 24: Comparison of the loading curves for the baseline geometry CU1 and the adapted round-nosed RD1. The values in the characterisation tensor of Fig. 15, obtained by identification procedure, show the

round splitter nose increases the collector losses (to a recognisable extent more in the preferred collector, leading to output terminal Y) and decreases the effectiveness of the interaction term s.

The availability of the splitter lengthening component used to set up the geometry CU2 was utilised to investigate the effect of the extremely short nozzle-to-splitter distance $\sigma = 2.59$ in the variant RD2 as presented in Fig. 25. In fact, no good efficiency in this extemporised variant was expected, considering the too small width of the main collector entrance – actually smaller than the nozzle exit width. The result of the measurements in Fig. 26 may be surprising only in finding out that the efficiency decrease relative to Fig. 22 is not very large, mere $\Delta\eta = 4\%$ spread uniformly over the whole of the no-spillover regime. In the jet-pumping regime the difference is larger – obviously due to decreased exposition of the outer side of the jet to the entrainment.

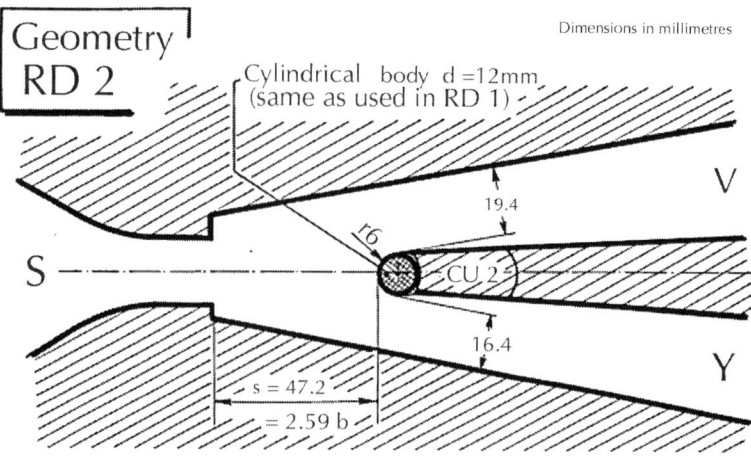

Figure 25: Another investigated adaptation of valve geometry, with rounded splitter at a short distance from the nozzle exit.

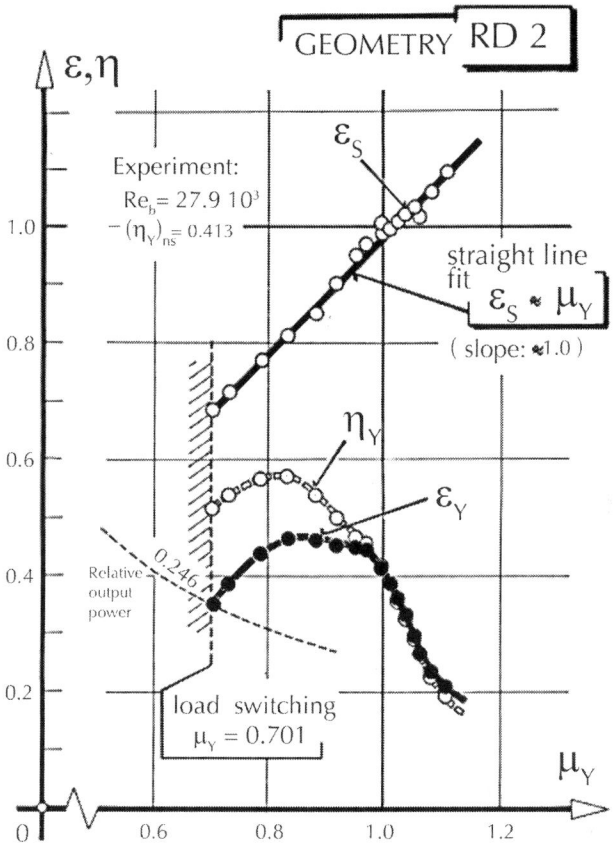

Figure 26: Output characteristic, in similarity co-ordinates, of the valve in the version from Fig. 25. Comparison with Fig. 22 shows that even a substantial change in the nozzle-to-splitter distance with the round-nosed splitter surprisingly does not lead to a significant change of the loading curve in the spillover regime.

WEDGE-SHAPED SPLITTER

A question posed quite often is why the splitter in author's valves is not of what seems to be the most obvious shape – simply a wedge with a sharp edge facing the nozzle exit. The present research provided also an opportunity to test this configuration. The small 6.5° apex angle of the splitter walls (Fig. 8) would place the sharp edge designed as their continuation too near to the nozzle; this is why the wedge geometry was tested with a shorter insert facing the exit with the 45° apex (Fig. 27). The lower inclined wall of the insert as shown in Fig. 27 also provided an improved opportunity for the jet entering the preferred collector (leading into Y) with the very small nozzle-to-splitter distance, which in this variant K1 was $\sigma = 3.19$. It has to be admitted that the measurement results, as presented in Fig. 28, are not bad – with the efficiency $\eta = 46.6\%$ they are better than in any of the variants discussed above. The position of the load-switching point is also acceptable. Like in the round (Fig. 23) and other cusp-less variants, the no-spillover efficiency value presented in Fig. 29 is remarkably constant and Reynolds-number independent.

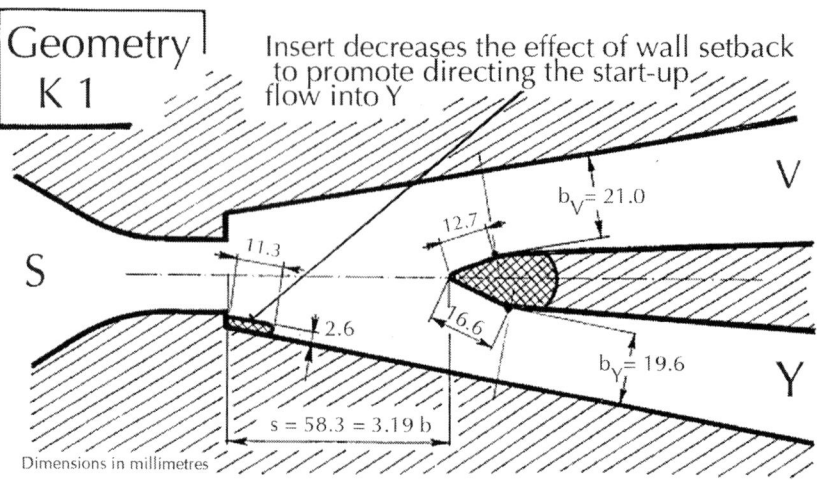

Figure 27: Investigated valve variant with sharp wedge splitter shape. To ensure the preference of the output Y, the setback of the main attachment wall had to be decreased – this could be done by a simple flat insert.

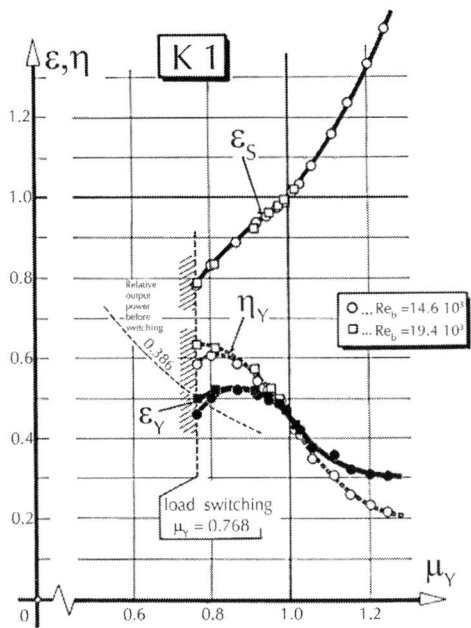

Figure 28: Experimental complete output characteristic of the valve in the wedge-shaped splitter version K1.

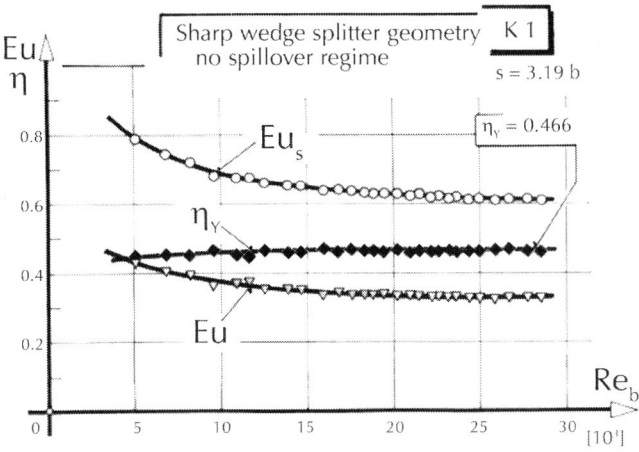

Figure 29: Euler numbers and efficiency for the variant from Fig. 27.

An interesting effect encountered in these experiments was the reluctance – not found in the earlier variants – of the jet to attach at starting the flow to the preferred attachment wall. An improvement in this aspect was achieved by placing at the beginning of this wall, as shown in Fig. 27, a simply shaped (rectangular cross-section) insert, which decreases the effective setback.

Later in version K2, this simple insert was replaced by a more properly shaped one, actually eliminating the setback of the attachment wall completely, as seen in Fig. 30. This, however, produced a too strong attachment, almost eliminating the load-switching effect, which was to be restored by placing a similarly faired insert (with a small setback, however) also on the opposite side. The loading behaviour at the two relatively small Reynolds-number conditions, as presented in Fig. 31, seemed to be quite good. A surprising fact, however, was later revealed by the tests over a wider Re range, presented in Fig. 32: the efficiency was found to decrease almost continuously with increasing Re and near the end of the range fell to a rather poor value.

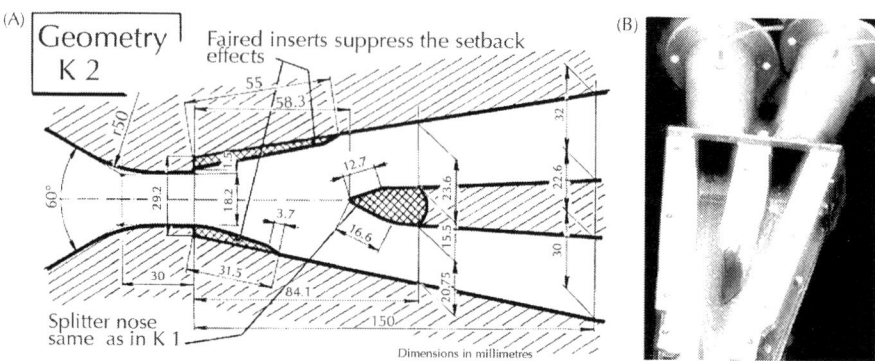

Figure 30: The significance of the attachment wall setbacks was tested in the valve geometry K2, with the wedge-shaped splitter differing from the case K1 by the setback corners filled on both sides with inserts.

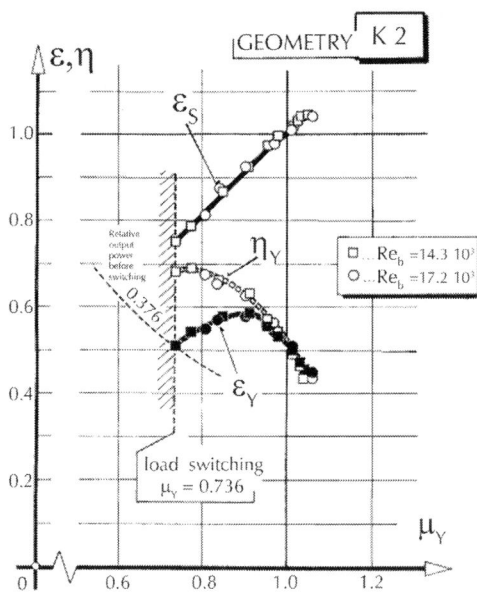

Figure 31: Output characteristic of the valve in the version K2 shown in Fig. 30. Surprisingly, a comparison with Fig. 28 shows the effect of the setbacks seems to be rather insignificant – just some improvement in efficiency.

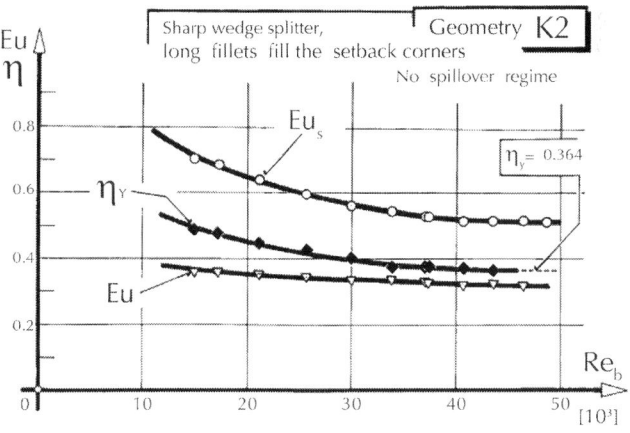

Figure 32: Euler numbers and efficiency for the valve variant from Fig. 30. As could be expected, there is a pronounced Reynolds-number dependence, apparently caused by increased role of wall-friction loss.

STRANGE CASE OF RE DEPENDENT SWITCHING

In one of the performed tests, the inserts filling the setback corners were exchanged, to test the magnitude of their influence. It was discovered that for the proper operation of the valve the splitter had to be placed into the body in an inverted position – with the result, as shown in Fig. 33, of the wedge tip positioned not on the nozzle axis (as was the case in Fig. 27 and Fig. 30) but with considerable eccentricity (more than 26% of the nozzle width towards the preferential attachment wall). The completely unexpected behaviour of this SK2 variant was the loss of similarity of loading characteristics: they became so strongly Re-dependent that the data points – as shown in Fig. 34 – failed to fall on a common curve. An even more surprising effect – so far never mentioned in literature – was this Re-dependence continuing so far that at a particular Reynolds number, in the investigated case at Re = 23.5 × 10^3, the available output energy fell to zero (Fig. 35).

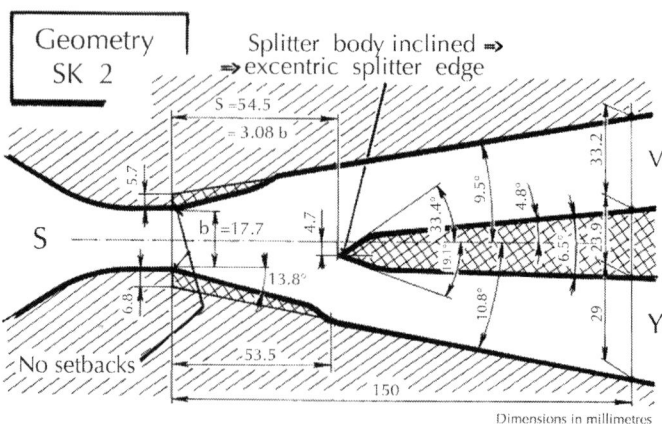

Figure 33: Valve geometry which demonstrated that the influence of the preferential attachment wall is not as small as could be inferred from comparison of Fig. 28 and Fig. 31. Here the preferential wall secures guiding the jet into a significantly narrower main collector entrance.

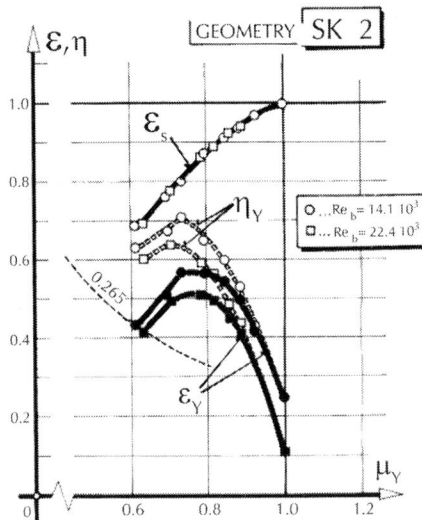

Figure 34: Output characteristic obtained experimentally with the geometry variant SK2. There is an unusual failure of the characteristics at different Reynolds number to form a single universal line in the similarity co-ordinates.

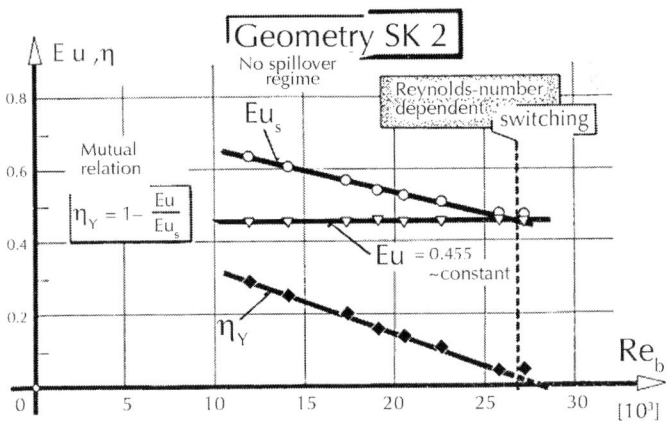

Figure 35: Euler numbers and efficiency dependence on Reynolds number for the valve variant from Fig. 33 explains the failure of the similarity transformation in the previous Fig. 34. Very special property found in

this variant is the existence of the limiting flow rate at which the Coanda effect fails to keep the jet attached in a situation different from the load-switching.

No doubt, such behaviour – an automatic change of character when a particular flow rate is reached – may find a number of interesting applications in chemical engineering.

SIMPLE SUPPLY CURVES IN THE OUTPUT CHARACTERISTIC

Character of the supply curves Eq. (9) as they were found in the output characteristics of the tested valves, may be quite complex, sometimes (e.g., in Fig. 12 and Fig. 18) with a sudden change of the behaviour in the no-spillover state $\mu_Y = 1$, while in other cases (e.g., in Fig. 22, Fig. 26 and Fig. 28) there is no change at all. The specific-energy difference between the supply inlet S and the vent terminal V is obviously mainly dependent on what happens in the interaction cavity of the valve: the loading cannot significantly influence the flow in the nozzle and also whatever changes are there in the vent collector and diffuser are certainly of little influence: the vent in the operating conditions of present interest is practically inactive, with no or very small flow. The large number of accumulated experimental results made possible extensive comparisons, from which it became obvious that a substantial difference in the character of these curves is between the behaviour of convex and concave bi-cuspid splitter shapes. This is obviously due to the different importance of the captive vortex in the interaction cavity [13].

The summary of data for the shapes not promoting this vortex is in Fig. 36. Very surprising – with respect to the typically quadratic or more generally non-linear characteristics of fluid flows – is the fact that Fig. 36 makes obvious: the supply line of the output characteristic in these cases is straight, with positive slope of unity. On the other hand, with most of the splitter shapes having the concave wall between the two cusps, the supply line (Fig. 37) is also linear with negative slope magnitude −1.0. Within these shapes there are more exceptional cases, one of them (actually the baseline case CU1) exhibiting slope magnitude −0.5.

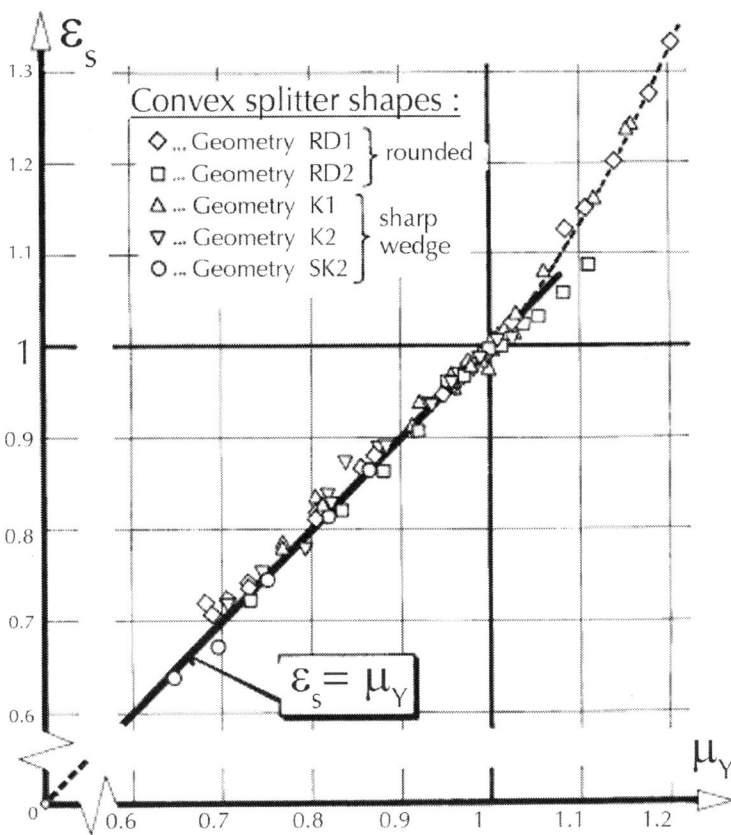

Figure 36: Measurements of the changes in the supply specific-energy drop with loading of the valve has shown a remarkable identity of the dependence for all geometries without the internal feedback: in the similarity co-ordinates the dependence is a simple linear proportionality.

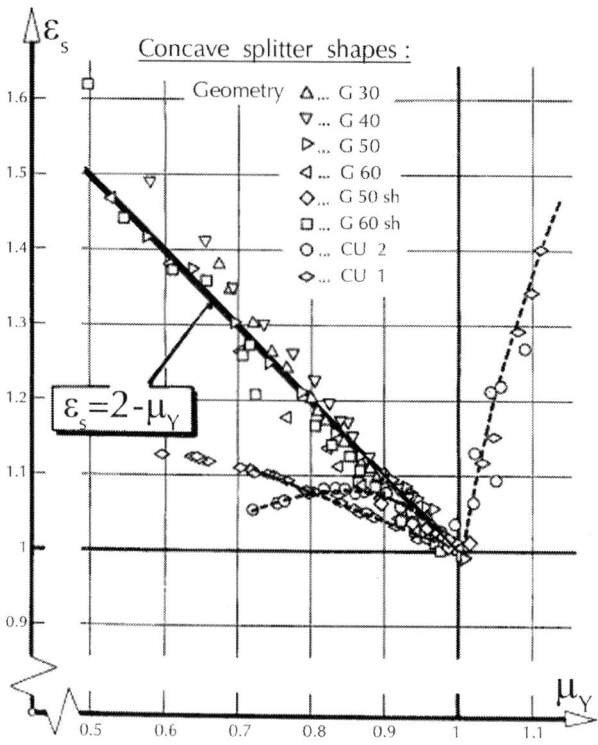

Figure 37: Similar plotting of the experimental results as in Fig. 36 for all investigated geometry variants with the bi-cuspid splitter noses generating the internal feedback.

CONCLUSIONS

Performed extensive experimental investigations of various alternative designs of a no-moving-part valve switching the flow into a different outlet, when experiencing a certain critical level of output loading, demonstrate that this is a device that deserves being better known in chemical engineering circles. The absence of any control circuits and of other sensitive and delicate instrumentation means such a valve may be easily – and inexpensively – made form materials making the valve resistant to extreme operating conditions (high temperature, vibration, or nuclear radiation) and capable of directing the flows of difficult to handle fluids – such as, e.g., hot gas, molten metals, or chemically aggressive liquids. The accumulated data provide useful information to designing such valves. The paper brings to attention several interesting new aspects, such as the characterisation by the third-order tensor, strange Re-dependent switching, and the linear +1 or −1 slope loading dependence of the supply energy difference in the similarity-transformed co-ordinates.

ACKNOWLEDGMENTS

The author gratefully acknowledges the financial support by research grant IAA200760705 and by the research plan AV0Z20760514 provided by the Grant Agency of the Academy of Sciences of the Czech Republic – as well as by the grant 101/07/1499 from the Grant Agency of the Czech Republic.

REFERENCES

1. V. Tesar, Extremely simple pressure regulator—computation studies, Chemical ˇ Engineering Journal 155 (2009) 361.
2. T. Scanlon, P. Wilson, G. Priestman, J. Tippetts, Development of a novel flow control device for limiting the efflux of air through a failed pipe, in: Proceedings of ASME Turbo Expo 2009, Paper GT2009, Orlando, Florida, USA, 2009.
3. V. Tesar, Fluidic control of molten metal flow, Acta Polytechnica: Journal of ˇ Advanced Engineering 43 (1) (2003) 15, ISSN 1210-2709.
4. V. Tesar, Fluidic valve for reactor regeneration flow switching, Chemical Engi- ˇ neering Research and Design, Part A 82 (A3) (2004) 1.
5. V. Tesar, Fluidic valves for variable-configuration gas treatment, Chemical ˇ Engineering Research and Design, Part A 83 (September (A9)) (2005) 1111– 1121.

6. V. Tesar, Fluidic control of reactor flow—pressure drop matching, Chemical ˇ Engineering Research and Design 87 (6) (2009) 817.
7. V. Tesar, et al., New ways of fluid flow control in automobiles: experience ˇ with exhaust gas after treatment control, in: Proceedings of World Automobile Congress, Paper No. F2000H192, Society of Automotive Engineers, Seoul, Korea, June 2000, p. 167, ISBN 89-85000-00-4 98550.
8. V. Tesar, Law for pressure loss in monolithic reactor matrices, in: Proceedings ˇ of Colloquium "Fluid Dynamics'98", Institute of Thermomechanics, Academy of Sciences of the Czech Republic, Prague, October 1998, p. 49, ISBN 80-85918- 45-5.
9. J.R. Tippetts, J.K. Royle, Design of flow control circuits involving bistable fluid amplifiers, Fluidics Quarterly 3 (4) (1971) 1.
10. J.R. Tippetts, Development Needs, NATO agency for aeronautical research and development AGARDograph 215 (1976).
11. N.K. Ibragimov, Elementary Lie Group Analysis and Ordinary Differential Equations, Wiley, New York, 1999.
12. V. Tesar, C.-H. Hung, W. Zimmerman, No-moving-part hybrid-synthetic jet ˇ actuator, Sensors and Actuators A 125 (2) (2006) 159.
13. V. Tesar, Mechanism of pressure recovery in jet-type actuators, Sensors and ˇ Actuators A: Physical 152 (2009) 182–191.
14. V. Tesar, Pressure-driven Microfluidics, Artech House, Boston, USA, 2007. ˇ
15. Fluidic circuits, in: W.B.J. Zimmerman (Ed.), Microfluidics: History, Theory, and Applications, Springer-Verlag, Wien/New York, 2006, pp. 255–304, CISM Courses and Lectures No. 466, ISBN-10-3-211-32994-3.
16. L.N. Trefethen, A.E. Trefethen, S.C. Reddy, T.A. Driscoll, Hydrodynamic stability without eigenvalues, Science 261 (5121) (1993) 578.

17. V. Tesar, W.B.J. Zimmerman, G. Regunath, Helical instability structures in ˇ swirling jets, Proceedings of the 8th Internat. Symp. on Fluid Control. Measurement, and Visualization FLUCOME 2005, Chengdu, China (2005).
18. V. Tesar, Characterisation of three-terminal fluidic elements and solution of ˇ bifurcated-flow circuits using the concept of equivalent dissipance, Journal of Fluid Control/Fluidics Quarterly 13 (1981) 55.

Citations

CHAPTER 1

Jianbo Yin and Xiaopeng Zhao, Electrorheology of nanofiber suspensions, doi: 10.1186/1556-276X-6-256.

CHAPTER 2

Francisco Solís-Pomar, Eduardo Martínez, Manuel F Meléndrez, and Eduardo Pérez-Tijerina,, Growth of Vertically Aligned Zno Nanorods Using Textured Zno Films, doi:10.1186/1556-276X-6-524

CHAPTER 3

Young-Jun Lee, Dae-Young Kim, Kap-Ho Lee, Moon-Hee Han, Kyoung-Soo Kang, Ki-Kwang Bae, and Jong-Hyeon Lee, Ammonium fluoride-activated synthesis of cubic -TaN nanoparticles at low temperatures, doi:10.1186/1556-276X-8-126.

CHAPTER 4

Nicolas Verplanck, Yannick Coffinier, Vincent Thomy, and Rabah Boukherroub, Wettability Switching Techniques on Superhydrophobic Surfaces, doi:10.1007/s11671-007-9102-4.

CHAPTER 5

Mark Lazari, Kevin M Quinn, Shane B Claggett, Jeffrey Collins, Gaurav J Shah, Henry E Herman, Brandon Maraglia, Michael E Phelps, Melissa D Moore, and R Michael van Dam, ELIXYS - a fully automated, three-reactor high-pressure radiosynthesizer for development and routine production of diverse PET tracers, doi:10.1186/2191-219X-3-52.

CHAPTER 6

Mainier, F., Fonseca, M., Tavares, S. and Pardal, J. (2013) Quality of Electroless Ni-P (Nickel-Phosphorus) Coatings Applied in Oil Production Equipment with Salinity. Journal of Materials Science and Chemical Engineering,1, 1-8. doi: 10.4236/msce.2013.16001.

CHAPTER 7

A. Cunha, M. Pacheco and J. Bergmann, "Influence of the Chemical Composition of Completion Fluids on the Propagation of Electromagnetic Waves within Oil Wells," Engineering, Vol. 4 No. 12A, 2012, pp. 966-971. doi: 10.4236/eng.2012.412A122.

CHAPTER 8

Václav Tesa, No-moving-part valve for automatic flow switching, Chemical Engineering Journal, Volume 162, Issue 1, 1 August 2010, Pages 278-295, ISSN 1385-8947, http://dx.doi.org/10.1016/j.cej.2010.04.028.

Index

A

Anisotropic structured carbon nanotubes (ACNTs) 100
Anodic aluminum oxide (AAO) 105
Atomic layer deposition (ALD) 46, 48
Atom transfer radical polymerization (ATRP) 105
Average height (AH) 52

C

Calcium and titanium precipitate (CTP) 10, 35

Carbon black (CB) 27
Carbon nanotubes (CNT) 124
Carbon nanotubes (CNTs) 2, 99
Carbonyl iron (CI) 17
Chemical vapor transport (CVT) 47

D

Differential scanning calorimetry (DSC) 73

E

Electron diffraction (ED) 6
Electrorheological (ER) 1

Electrostatic field strength (EFS) 28
ElectroWetting On Dielectric (EWOD) 114
Energy dispersive X-ray spectroscopy (EDS) 50

F

Fluid catalytic cracking (FCC) 16

H

Hexamethylenetetramine (HMT) 49
High-performance liquid chromatography (HPLC) 146

L

Lower critical solubility temperature (LCST) 105

M

Magnetorheological (MR) 2

O

Outside diameter (OD) 152

P

Positron-emission tomography (PET) 137
Pulsed laser deposition (PLD) 47

Q

Quaternary methylammonium (QMA) 149

S

Scanning electron microscopy (SEM) 52, 126
Selected area electron diffraction (SAED) 79
Solid–liquid–solid (SLS) 112

T

TEM propagation 190
Thermogravimetric analysis (TGA) 73
Transmission electron microscopy (TEM) 6
Transverse electromagnetic mode (TEM) 193

U

University of Los Angeles (UCLA) 121

V

Vapor-grown carbon nanofibers (VGCFs) 22
Vapor–liquid–solid (VLS) 126

X

X-ray diffraction (XRD) 49, 71